"With an encyclopaedic knowledge of transhumanism and a deep philosophical grounding, especially in Nietzschean thought, Stefan Sorgner tackles some of the most challenging ethical issues currently discussed, including gene editing, digital data collection, and life extension, with uncommon good sense and incisive conclusions. This study is one of the most detailed and comprehensive analyses available today. Highly recommended for anyone interested in transhumanist/posthumanist ideas and in these issues generally."
N. Katherine Hayles, University of California, Los Angeles

"An eye-opening, wide-ranging and all-inclusive study of transhumanism. Sorgner's account avoids both the utopian trap and the bogeyman spectre. He makes a compelling case for placing ourselves on the transhuman spectrum. How we continue to use technologies is in our hands. Sorgner's book is both a comprehensive introduction to transhumanist thought and a clear-sighted vision for its future realisation."
Julian Savulescu, University of Oxford

WE HAVE ALWAYS BEEN CYBORGS

Digital Data, Gene Technologies, and an Ethics of Transhumanism

Stefan Lorenz Sorgner

BRISTOL
UNIVERSITY
PRESS

First published in Great Britain in 2023 by

Bristol University Press
University of Bristol
1-9 Old Park Hill
Bristol
BS2 8BB
UK
t: +44 (0)117 374 6645
e: bup-info@bristol.ac.uk

Details of international sales and distribution partners are available at bristoluniversitypress.co.uk

Chapter 2.1 *Transhumanism without Mind Uploading and Immortality : this chapter appears in Analyzing Future Applications of AI, Sensors, and Robotics in Society* (pp. 284– 291) edited by Musiolik, T.H., & Cheok, A.D. © 2021, IGI Global, www.igi- global.com . Reprinted by permission of the publisher.

Chapter 2.2.11 *Glocalization and the War for Digital Data : this section appears in Journal of Posthuman Studies* 4(2), (2020), 'Editor's Note', Stefan Lorenz Sorgner © 2020, The Pennsylvania State University Press. This article is used by permission of The Pennsylvania State University Press.

British Library Cataloguing in Publication Data
A catalogue record for this book is available from the British Library

ISBN 978-1-5292-1920-3 hardcover
ISBN 978-1-5292-1921-0 paperback
ISBN 978-1-5292-1922-7 ePub
ISBN 978-1-5292-1923-4 ePdf

Cover design: blu inc, Bristol
Front cover image: istock-1150039017

Bristol University Press uses environmentally responsible print partners.

Printed in Great Britain by CPI Group (UK) Ltd, Croydon, CR0 4YY

FSC
www.fsc.org
MIX
Paper | Supporting responsible forestry
FSC® C013604

Contents

List of Abbreviations   vi
Acknowledgements   vii

1   Transhumanism: In a Nutshell   1
2   On a Silicon-based Transhumanism   22
3   On a Carbon-based Transhumanism   61
4   A Fictive Ethics   109
5   The End as a New Beginning   185

Notes   188
References   198
Index   209

List of Abbreviations

AI	Artificial intelligence
IEET	Institute for Ethics and Emerging Technologies
IVF	In vitro fertilization
KSA	Nietzsche, F. (1967ff) *Sämtliche Werke. Kritische Studienausgabe in 15 Bänden.* Edited by G. Colli and M. Montinari. München/New York: Deutscher Taschenbuch Verlag.
MIT	Massachusetts Institute of Technology
PA	Procreative autonomy
PB	Procreative beneficence
PGD	Preimplantation genetic diagnosis
PGS	Preimplantation genetic screening
RFID	Radio-frequency identification
WP	Nietzsche, F. (1968) *The Will to Power.* Trans. Walter Kaufmann and R.J. Hollingdale. Weidenfeld and Nicolson, London.
WTA	World Transhumanist Association

Acknowledgements

I wish to thank my Research Assistants of John Cabot University (JCU) Rome for their support for getting this manuscript ready for publication: Selma Coleman, Moya Seneb, Moustafa Tlass, Megan Dhlamini, Benedetta Grilli, and Francesca Dalmazzo. I particularly wish to highlight the efforts for the polishing of this manuscript of my JCU Research Assistants Ihsan Baris Gedizlioglu, Daniela Movileanu, and Chryssi Soteriades. Furthermore, I am grateful for the exchanges I have had, and comments I received, from Dr Markus Peuckert, Dr Pascal Henke, and Prof Thomas Govero.

Transhumanism: In a Nutshell[1]

This book on 'We Have Always Been Cyborgs' is structured as follows. Chapter 1 will be a general introduction to transhumanism. I will critically analyse the wide range of digital developments relevant for transhumanism in Chapter 2, 'Silicon-based Transhumanism'; for example, mind uploading and cyborgization. In Chapter 3, 'On a Carbon-based Transhumanism', my focus will be on the wide range of gene technologies which are central for transhumanism, that is, (1) Nietzsche and recent debates on transhumanism and eugenics; (2) critical reflections on moral bioenhancement; (3) gene modification; (4) gene selection after in vitro fertilization (IVF) and preimplantation genetic diagnosis (PGD). In Chapter 4 the main ethical discussions concerning transhumanism will be summarized and I will present my own fictive ethical stance, that is, (1) virtue ethics; (2) the question of the good life; (3) personhood and what is morally right; (4) transhumanism and utopia; (5) transhumanism, immortality and the meaning of life. By this means my key thought that we have always been cyborgs in the continual process of self-overcoming, will unfold itself in various dimensions. To begin with, however, an informed understanding of transhumanism needs to be presented.

Transhumanism is the 'world's most dangerous idea'. This is at least Francis Fukuyama's judgement concerning this cultural and philosophical movement, which he stated in the magazine *Foreign Policy* (Fukuyama 2004, 42–43). Transhumanism is a cultural movement which affirms the use of techniques to increase the likelihood that human beings manage to transcend the boundaries of their current existence. It is in our interest to take evolution into our own hands. Thereby, we will increase the likelihood of our living a good life as well as that of not becoming extinct.

Transhumanism has slowly increased in significance since 1951, when the term was first coined by Julian Huxley in his article 'Knowledge, Morality, and Destiny'. Then, he described transhumanism as follows: 'Such a broad philosophy might perhaps best be called, not Humanism, because that has certain unsatisfactory connotations, but Transhumanism. It is the idea of humanity attempting to overcome its limitations and to arrive at fuller fruition; it is the realization that both individual and social developments are processes of self-transformation' (Huxley 1951, 139). I regard this formulation still as the best possible definition of transhumanism.[2]

The concluding chapter of Julian Huxley's book *New Bottles for New Wine*, published in 1957, is entitled 'Evolutionary Humanism'. The relationship between evolutionary humanism, represented today by the Giordano Bruno Foundation, and contemporary transhumanism must still be clarified more precisely. There seems to be a structural analogy between transhumanism and evolutionary humanism which needs to be considered when clarifying the relationship between humanism and transhumanism, and also between traditional humanism and evolutionary humanism.

Julian Huxley had a brother who is at least as well known as he himself, Aldous Huxley. Between Julian Huxley's affirmative considerations concerning the impacts of technologies and those of his brother, the author of the critical novel *Brave New World*, there are significant tensions in terms of content. Julian Huxley also shares his fundamental evolutionary approach with his grandfather Thomas Henry Huxley, who distinguished himself as Darwin's supporter. He was known as Darwin's bulldog. Julian Huxley's half-brother, Andrew Fielding Huxley, was also active as a natural scientist. He was a university professor of biology in London and won the Nobel Prize in Physiology or Medicine, but is currently less well known than the other family members already mentioned. Julian Huxley was a university professor in London, too. In addition, he was the first general director of UNESCO who made a significant contribution to the Universal Declaration of Human Rights, and was on the board of the British Eugenics Society.

A close friend of Julian Huxley was the catholic evolutionary thinker Teilhard de Chardin, who used the word 'transhumanising' in *The Future of Man* (De Chardin 1959, 251). The reflections by this Jesuit priest are still of great relevance for considering potential religious aspects of transhumanism and for further clarifications concerning the relationship between Christianity and transhumanism.

In the time period 1969 to 1972 there were six manned US landings on the Moon, which significantly revived interest in this way of thinking. It was then that the notions of the 'post-' and the 'transhuman' in a transhumanist sense were coined. Both concepts definitely show up in the article 'Transhumans-2000' by F.M. Esfandiary (1974), who changed his name to FM-2030 to stress the contingency of naming conventions and to highlight the relevance of a prolonged life span, as 2030 would have been the year of his 100th birthday: 'On our way beyond animal beyond transhuman – to a post-human dimension' (Esfandiary 1974, 298). In his

1973 book *Up-Wingers* he had already talked about a 'post-animal/human stage' (Esfandiary 1973, 170). The notion 'superman' is prominent in the seminal book *Man into Superman* by R.C.W. Ettinger (1972). It is the notion of the superman which relates transhumanist reflections to the philosophy of Friedrich Nietzsche. A detailed debate (see Tuncel 2017) on the complex relationship between these two ways of thinking was initialized in 2009 by my article 'Nietzsche, the Overhuman, and Transhumanism' (Sorgner 2010b, 2016a, 2018c, 2019b). Most leading transhumanist philosophers, however, belong to the tradition of Anglo-American analytical philosophy, which is one reason why there is a widely shared hesitation within the continental philosophical tradition to seriously engage with transhumanist thinking (More and Vita-More 2013).

Further, significant ancestors of transhumanist thinking can be found within Russian cosmism and Russian science fiction. It would be anachronistic to refer to any thinker before 1951 as transhumanist, yet many structural analogies and parallels can be found between their reflections and the realm of topics which is usually covered by transhumanist thinkers, activists, and artists. Nikolai Fyodorovich Fyodorov is the most noteworthy thinker among the Russian cosmists concerning the similarities to transhumanist reflections. In contrast to Nietzsche, who was hostile towards religions, Fyodorov was a devout, church-going Orthodox Christian, which also influenced his futuristic ideas. At least as fascinating as the cosmists were Russian science fiction writers. Alexander Romanovich Belyaev's books *Professor Dowell's Head*, *Amphibian Man*, *Ariel*, and *The Air Seller* are particularly relevant. The novel *We* by Yevgeny Ivanovich Zamyatin is also noteworthy in this context, as well as the brothers Arkady Natanovich Strugatsky and Boris Natanovich Strugatsky, who developed the complex world of 'Noon' in which several of their novels take place, and which was named after the first novel of their series: *Noon: 22nd Century*.

Transhumanism as a cultural movement developed further during the 1980s with a close friend of FM-2030, Natasha Vita-More. In 1982 she published the 'Transhuman Manifesto', which preceded the first version of the 'Transhumanist Arts Statement' (1992), of which a revised version was republished in 2003. In 2013 she summarized her central insights on these issues in the article 'Aesthetics: Bringing the Arts and Design into the Discussion of Transhumanism' (Vita-More 2013, 18–27). Like FM-2030, who was born as Fereidoun M. Esfandiary, she changed her name to highlight the contingency of naming. Natasha Vita-More was born as Nancie Clark, and her husband Max More as Max T. O'Connor. Max More, Natasha Vita-More, and FM-2030 were particularly relevant in promoting transhumanism at the beginning of the 1990s. Max More's essay 'Transhumanism. Toward a Futurist Philosophy' (More 1990, 6–12) and FM-2030's book *Are You a*

Transhuman (FM-2030, 1989) were particularly important writings from this period. Later during the 1990s some transhumanists realized that a more formally organized structure was needed to increase the cultural impact of transhumanism. Consequently, the World Transhumanist Association (WTA) was founded by Nick Bostrom and David Pearce in 1998. Nick Bostrom is particularly famous for the paper 'Are You Living in a Computer Simulation?' (Bostrom 2003, 243–255). David Pearce is widely known for having authored the 'Hedonist Imperative' (Pearce 1995). The WTA also established the conference series TransVision and founded the *Journal of Transhumanism* to promote transhumanism as a legitimate field of academic research.

Still, the WTA did not realize its expected academic impact. Hence, Nick Bostrom and James Hughes established the techno-progressive think-tank Institute for Ethics and Emerging Technologies (IEET) in 2004 and integrated the *Journal of Transhumanism* into the structures of this organization. Thereby, the peer-reviewed, academic, open access, online-only journal was renamed as the *Journal of Evolution and Technologies*. In order to generate an increased impact and to reduce the fear resulting from the word 'transhumanism', the WTA was rebranded in 2008, being renamed as 'Humanity Plus' and having cultural activism as its focus, whereas the focus of the IEET is academic and it is open not only to transhumanists but also to other techno-progressive thinkers, like bioliberals. You can be a bioliberal and regard it as morally obligatory to select specific fertilized eggs after IVF and preimplantation diagnosis for implantation without regarding the coming about of a posthuman as desirable. This is a position which is held by Julian Savulescu, who is a prime example of a philosopher and bioethicist who is close to transhumanists' thinking but does not explicitly associate himself with this movement. On the other hand, it can also be the case that you reject the notion of any moral obligation, while regarding it as individually desirable to overcome current human limitations by means of technologies, which is the distinguishing feature of transhumanists.

A further opening towards academic debate took place with the realization of the Beyond Humanism Conference Series in 2009, for which I have been primarily responsible. Its goal is to promote academic exchange between humanist, transhumanist, and critical posthumanist scholars. This is also the dedicated aim of the book series *Beyond Humanism: Trans- and Posthumanism*, which I established in 2011. To further promote the goal of an academic engagement with the great variety of discourses, James Hughes, Sangkyu Shin, and I established the world's first double-blind and peer-reviewed academic journal dedicated to the posthuman, the *Journal of Posthuman Studies*, which has been published in print as well as online by Penn State University Press since 2017. Since 2019, the world's oldest publishing house, Schwabe Verlag (founded in 1488), which is deeply rooted in the humanist tradition, has embraced the challenges related to emerging technologies and has published a high-class book series entitled *Posthuman Studies* under my general editorship.

1.1 Philosophical issues

A main philosophical issue with which I will be less concerned here is that of transhumanist politics. The media widely identifies transhumanists with cold-hearted, blood-sucking, Silicon Valley billionaires who fight for libertarianism. This description is one sided. There are libertarian transhumanists like Peter Thiel or Zoltan Istvan, but the majority of intellectual transhumanists are in favour of a social-democratic version of transhumanism, as it is upheld by most members of the IEET. Particularly noteworthy concerning transhumanist politics are the activities of Zoltan Istvan, who founded the US Transhumanist Party and who ran for the US presidency in 2016. In the novel *Transhumanist Wager* (2013), Istvan explains many of his transhumanist positions. Due to his libertarian sympathies, his activism was not universally approved of by transhumanists. A complex social-democratic version of a transhumanist politics had been published in the book *Citizen Cyborg* by James Hughes (2004). This monograph not only presents a strong political position but also serves as an excellent introduction to transhumanism in general.

No serious transhumanist rejects liberalism. All affirm some liberal political stance, but the range of possible positions is enormous. This is also the reason why the right to morphological freedom plays such a central role in transhumanism. The concept was originally coined by Max More in his 1993 article 'Technological Self-Transformation: Expanding Personal Extropy' (More 1993, 15–24). A particularly strong case for morphological freedom was presented by Anders Sandberg in his article 'Morphological Freedom – Why We Not Just Want It, but Need It' (Sandberg 2013, 56–64). This right includes not only the self-ownership of one's body but also the right to modify it according to one's own wishes.

This right can be explained further by reference to a utilitarian foundation, as most academic transhumanists belong to the Anglo-American analytic philosophical tradition, where utilitarianism is dominant. There are some particularly strong transhumanist scholars in the disciplines of applied ethics and the philosophy of mind. Academic contributions by these were responsible for the recognition of transhumanism in academia as an approach which deserves to be taken seriously, from the early 2000s onwards. This does not mean, however, that there are no transhumanists who argue on the basis of a different ethical theory. James Hughes used to be a Buddhist monk, and many of his arguments are founded on a virtue-ethical approach, for example when he stresses the need to technologically enhance mindfulness. I argue for a hermeneutical ethical position, which is strongly influenced by Nietzschean reflections: a liberal ethics of fictive autonomy.

Depending on the ethical foundation, different concepts of the good are being upheld by transhumanists, even though in the mainstream media transhumanism is usually presented as if all transhumanists wish to become immortal by turning into a Renaissance genius, whereby the male version can be identified with Superman on Viagra, and the female ideal is best described as Wonder Woman with Botox. It is not the case that no transhumanist upholds these ideals, but it needs to be pointed out that there is a much greater diversity concerning the concepts of the good which are being affirmed than has widely been acknowledged. I will discuss this issue in detail in Chapter 4, section 2.

The main aspect of the concept of the good which is shared among many transhumanists is the prolongation of the human health span. The health span must not be identified with the life span, the quantity of years we are alive. What most people are interested in is the prolongation of healthy years of their lives. This is what the health span stands for. One academic discussion among transhumanists is about the status of this claim. Should the normative claim be upheld that all human beings ought to identify any prolongation of the health span with an increase in the likelihood of living a good life? Is it the case that the health span is valid for most human beings, but that it is not a universally valid normative ethical position? Does the affirmation of the prolongation of the human health span imply the position that immortality is being aimed for, or not? Is transhumanism a utopian enterprise, or not? Should ageing be seen as a disease (de Grey and Rae 2010)? Is cryonics a possible way to increase one's health span? Is mind uploading the best possible option for doing so, or should we create human–animal hybrids to move beyond the 122-year age limit which seems to apply to currently living human beings? All of these issues are being discussed in academic exchanges by transhumanists. I analyse them in detail in this book.

My own take on these issues is that most human beings indeed identify an increase of the health span with a higher likelihood of living a good life. Yet, neither cryonics nor mind uploading seem to be the most promising techniques for promoting this goal. This being the case, it needs to be pointed out that immortality is not a realistic option. We cannot even conceptualize the notion of immortality on the basis of a naturalist understanding of the world. Immortality implies either that humans cannot or that they must not die. Both options are absurd, if we think of the world on a naturalist basis. In this case, the universe began with the Big Bang. It has expanded since then, and eventually the expansion process will slow down so that the entire universe will come to a standstill. A different theory is that at a certain stage the expansion process will become revised into a contraction process that will end up in a cosmological singularity, a point of immense density. In both cases there does not seem to be a realistic option for human beings to survive. Yet this is what ought to be possible if immortality were an option. Hence, it is clear that immortality in its literal sense cannot be a realistic option for most transhumanists (Sorgner 2018b). Yet many transhumanists use, mention, and deal with the notion of immortality (Rothblatt 2015). If the person in question is a serious thinker, the concept of immortality ought to be used only in a rhetorical sense. It can be employed to create media attention, and as a means to generate a public awareness of a widely shared aspect of a good life, namely the prolongation of the human health span. Immortality in this case merely serves as an advertising tool for generating attention for an enormously important aspect of the good life, namely for stressing the relevance of increasing our health span. No serious transhumanist ought to affirm immortality in its literal sense as a realistic option. Furthermore, I defend a non-utopian version of transhumanism, as I regard utopias as extremely dangerous (Sorgner 2018c). Thereby, the present becomes sacrificed for a future which most probably will never be actualized. Yet I am aware that this attitude is not universally shared among transhumanists (see 'Letter from Utopia', Bostrom 2008, 1–7). The issue of utopia is another important academic topic concerning transhumanism (see Hauskeller 2012, 39–74).

The question concerning appropriate techniques is an important one. Which enhancement techniques are particularly relevant for bringing about a posthuman? One enhancement technique, which was already in place in antiquity, is surgery. By means of surgery one can change the shape of one's eyes or one's nose, enlarge one's penis, reconstruct the hymen, create bigger breasts, or have a sex reassignment. All of these interventions are widely used and popular in all parts of the world. This issue is particularly tricky when a client wishes to have his healthy leg removed because it does not belong to himself, according to his own self-understanding. A surgeon in the UK actually performed such surgery on two of his clients who had healthy legs.[3] Hymen reconstruction or hymenoplasty is particularly sought after by young Muslim women in Germany and the Netherlands, when they are no longer virgins and fear not getting married.[4]

Another long-established technique is pharmacological enhancement. Blood doping, pain killers, cognitive enhancers, anabolic drugs, love drugs, and many other types of pharmaceutical methods can be used to enhance human capacities. Each specific moral issue deserves a separate treatment. Oxytocin, MDMA, or Selective Serotonin Reuptake Inhibitors (SSRIs) can be taken to promote the mutual feeling of love. Microdosing LSD is the act of consuming sub-perceptual amounts of psychedelics, which has become popular in Silicon Valley since 2011[5] in order to promote creativity, gain energy, treat depression or anxiety, or even promote a spiritual awakening process. It is often the case that drugs which were used for treating diseases can also be employed to promote a certain capacity in healthy clients. Studies indicate that one in three US university student takes Adderall so as to be better prepared for an exam.[6]

However, for promoting the likelihood of bringing about a new human, a posthuman, other interventions are even more promising. Gene technologies, cyborg technologies, as well as digital technologies, seem to be the most important ones within transhumanists debates. The issues can be dealt with in an even more complex manner if we include altered external settings, like total surveillance systems, smart cities, the Internet of Things, the internet of bodily things in upgraded humans, and autonomous cars. Furthermore, an additional topic has been prominent in recent years: moral bioenhancement; there has been a popular and intense debate during the 2010s on various aspects of this topic (Sorgner 2016d). Traditionally, enhancements have been discussed which promote qualities which are widely identified as advantages, so that the likelihood of a person's leading a better life can be promoted. What about morality? Can moral bioenhancements at all be conceptualized? Should moral bioenhancement procedures become lawful, if they were available? Who would be interested in promoting their likelihood of acting morally? Would or should Catholics be interested in moral bioenhancement procedures to increase the likelihood of being rewarded with a fulfilled afterlife?

I regard gene technologies as well as cyborg technologies as the most promising means for expanding human boundaries (Sorgner 2018b). Within gene technologies, gene modifications as well as gene selection have to be distinguished. Gene modifications are particularly promising due to the development of CRISPR-Cas9, as well as other genome-editing methods. Gene selection can occur during the process of an IVF if several fertilized eggs are available, as is the case in the UK, but not in Germany. It is possible to analyse the fertilized eggs genetically and decide which ones should be implanted. In the meantime, even artificial intelligence (AI) can be a powerful tool and make this analysis. My own take on these issues is that it is advisable to seek established processes which are structurally analogous to new ones for developing an initial ethical take on new moral challenges. Then, it is possible to transfer the established moral norms to the new procedure. This approach has the following implications concerning parents' application of genetic technologies on their offspring. There is a structural analogy between parents genetically modifying their children and traditionally educating them (Sorgner 2015b, 31–48). Hence, the same moral standards ought to apply to both procedures. In both cases parents make decisions to alter their children in order to increase the likelihood of them living a good life. Not all genetic modifications are morally legitimate, in the same way as not all types education are morally appropriate. Some procedures which count as child abuse also ought to be banned and penalized. However, this explanation also reveals that in principle it can be morally appropriate for parents to genetically enhance their children. Whether certain procedures also ought to become lawful is a crucial issue. Certain vaccinations are legally obligatory in the US, and vaccinations are bioenhancements too. The other controversial case concerning gene technologies is that of selecting fertilized eggs after IVF and PGD. Concerning this procedure, the same standards ought to apply as in the case of someone looking for a partner for reproductive reasons, I argue, as selecting a partner for reproductive reasons is structurally analogous to selecting a fertilized egg after IVF and PGD (Sorgner 2014b, 199–212). Another promising genetic technique is that of gene analysis, in particular, if information techniques are combined with gene techniques. Big data analysis of genes is a necessary prerequisite for the other genetic techniques just mentioned, as it is necessary for gaining information concerning correlations between genes and gene clusters and traits, like qualities, diseases, or reactions towards drugs.

Cyborg technologies are a further flourishing field, and a continually increasing amount of scientific examples reveal what it can mean; for example, RFID (radio-frequency identification) chips may be implanted into different parts of our bodies (Sorgner 2019c). Our turning into cyborgs is a development which has taken place since we became *Homo sapiens*. Acquiring a language is the first human upgrade. Parents turn us into cyborgs by teaching us a language. The term 'cyborg' means cybernetic organism. The word *cyber* comes from the Ancient Greek κυβερνήτης, which means helmsman or pilot. So a cyborg is a governed, a steered organism. By being upgraded with language, we are being steered by our parents. Education is all about steering children. Vaccinations are a further means of doing so. Recently, RFID chips are becoming integrated into our bodies[7]; for example, into our front teeth to measure what we eat and drink (developed by scientists from Tufts University)[8]. This information then can be sent to our smartphone so that we can receive further advice concerning our diet. The most promising technique in this context is that of predictive maintenance. So far, predictive maintenance is being used in machines. Sensors tell us if a certain part of a machine will most probably fail within the next six months. We receive this information at a stage when the part is still fully functional. The same procedure can be applied in humans. Predictive maintenance of the human body can be used to adapt one's diet to avoid an increase in our blood sugar level. It is this technique of predictive maintenance which can enable us to radically increase the likelihood of an increased health span.

Quite a few transhumanists regard the option of mind uploading as the most promising enhancement technique, whereby we would turn our carbon-based personality into a silicon-based one: we digitize our personality. Bostrom's simulation argument (Bostrom 2003) is an interesting thought experiment which presupposes mind uploading. It demands careful consideration. However, from my perspective, its pragmatic relevance is comparable to the academic question of how many angels may dance on the head of a pin, which used to be discussed during the Middle Ages by the scholastics. Both arguments are interesting and make sense from the culturally dominant paradigms of each time. However, they are of no relevance whatsoever concerning any practical issues. My main reason why I regard the possibility of mind uploading as highly problematic is that we have no reason to believe that life can exist on a silicon basis. Computer viruses are self-replicating entities, but they do not have a metabolism for gaining energy, which is a central feature for other living entities. We have no indication for believing that digital life can even be possible As uploaded personalities we would still want to be alive, and being alive also seems to be a necessary prerequisite for having consciousness or self-consciousness, too. All of these reflections indicate that mind uploading is a highly dubitable procedure. I cannot exclude that it will eventually be possible, but our current scientific basis does not provide us with a strong reason for regarding it as a likely option. This is the reason why I compare it to the medieval angels argument. Kurzweil's reflection on mind uploading turns this procedure into a quasi-religious claim which I find highly problematic as well as implausible (Kurzweil/Grossmann 2011). This does not mean that there has to be a tension between a traditional type of religiosity and transhumanism, but the quasi-religious concepts developed so far have often been wild speculations. Still, it needs to be acknowledged that research concerning the relationship between religious experiences and technical processes, which supports such experiences, as well as other religious issues, deserve further studies in the future, such as the relationship between non-dualist anthropology, and the possibility of technologically induced religious experiences.

A further topic with which I will be concerned in a separate forthcoming monograph is that of transhumanist arts. Series like Black Mirror, West World, and Electric Dreams have been highly relevant in sparking a wide-ranging interest in transhumanism. The Hollywood movie *Transcendence* with Johnny Depp, the series *Big Bang Theory*, as well as the novel *Inferno* by Dan Brown have reached an audience of millions of people. In all of these works, transhumanism is addressed either implicitly or explicitly. Comics, movies, science fiction, as well as computer games are full of references to transhumanist ideas. In addition, the fields of body art, performance art, new media art, and bioart also deserve to be studied concerning their transhumanist implications. The works of the following artists deserve particular attention in this context: Stelarc, Orlan, Eyeborg Neil Harbison, Moon Ribas, Eduardo Kac, and Patricia Piccinini.

1.2 Transhumanism as nihilistic, positive pessimism

In this chapter I show why transhumanism can be identified with a nihilistic, positive pessimism. However, I am not claiming that all transhumanists ought to affirm and agree with this description. A survey provides significant evidence that the majority of transhumanists self-identify with a type of naturalism, atheism, or secularism.[9] I take this as a central starting point for my analysis. In section 1.2.1 I provide some reasons for claiming that a properly thought-through naturalism leads to a philosophical pessimism (Sorgner 2007), whereby I draw upon the usage of the term pessimism which goes back to Schopenhauer's self-description of his own philosophical account (Auweele 2017). Given the reflections on pessimism 1.2.1, section 1.2.2 is dedicated to the analysis how it can be possible to be both pessimistic and positive, what positivity stands for, what the underlying lines of thought for affirming positivity are, and why transhumanists can be identified with having a positive attitude. By drawing upon the former reflections, section 1.2.3 specifies the meanings of two different types of nihilism, alethic nihilism and ethical nihilism (Sorgner 2010a), which both can be seen as a plausible inference from a naturalistic ontological starting point, whereby this starting point is not a proper ontology but, rather, an ontology of continual becoming. Alethic nihilism implies that all philosophical judgements are interpretations, whereby the concept of interpretation does not imply that it has to be false, but merely that it can be false. Ethical nihilism, on the other hand, affirms that any non-formal judgement concerning the good life is plausible. Both concepts will be spelled out in more detail as part of these reflections. Furthermore, it also needs to be pointed out at the outset that a proper naturalism, if it becomes conceptualized philosophically, cannot be a naïve naturalism which is merely a new kind of atheistic essentialism. This will be discussed in more detail in section 1.2.3, when I explain why a philosophically informed naturalism ought to imply an alethic nihilism, with the plausibility of an alethic nihilism being best explainable on the basis of a non-naïve naturalism (Sorgner 2017a).

1.2.1 Pessimism

At the beginning of the 19th century more than 90% of the world's population had to deal with a daily struggle for survival.[10] They lived in absolute poverty. They did not have sufficient food, clean water, or suitable housing. They suffered enormously when it was cold in the winter or too hot in the summer. Not having the most basic needs for survival is connected with an enormous amount of suffering. However, even if you belong to the lucky few who do not have to struggle to survive at the moment, there are an enormous amount of challenges that you can be confronted with: earthquakes, volcanic eruptions, tsunamis, diseases, pandemics, wars, and the fear of losing your social privilege and being among the masses who have to struggle to survive. No matter whether you are rich or poor, you are confronted with continuous suffering. Whatever you do involves the overcoming of obstacles. You are stuck in a traffic jam, in a waiting room, or with your colleagues in a meeting. The moments of pleasure are rare, and usually last for only a short period of time. You long for an espresso, but it takes less than 20 seconds to drink it. You long to finish writing an article, but once it has been accepted the next deadline had to be met. You long for some exciting sexual highs, but they are over in a few fleeting moments. Most of the time we are struggling, fighting, and striving, but the moments of fulfilment are few and do not last long.

Dukkha is the first of the four noble truths of Buddhism, and it stands for the permanent painfulness of our lives (Agganyani 2013). It also corresponds with the basic analysis of existence put forward in Schopenhauer's philosophy (Ryan 2017). Schopenhauer referred to such an analysis of existence as pessimism. Nietzsche agrees with Schopenhauer's basic analysis of existence. However, their responses to the permanent suffering differ. While Schopenhauer agrees with a Buddhist mindset and stresses that a good life involves the long-term overcoming of suffering by means of asceticism (art merely provides a short-term relief), Nietzsche was in favour of affirming suffering in order to reach a moment which enabled him to justify all of his existence, *amor fati*, and make him love his fate (Urpeth 1999). Like Heraclitus, Nietzsche sees the world as continual becoming in all respects. Once you doubt the existence of Platonic forms, an unchanging human nature, and an immaterial soul, free will or reason, what remains is such a world of continual becoming (Sorgner 2007, 2010a).

According to the previously mentioned questionnaire, most transhumanists are naturalists, sceptics, and atheists. There are many types of naturalism, but naturalism usually implies merely the existence of entities which can be accessed empirically. Not too many scholars are both believers in a monotheistic religion and transhumanists. However, Buddhism is a religion which is affirmed widely among transhumanists, with James Hughes being the best-known example of a transhumanist Buddhist[11] (his next monograph is provisionally entitled 'Cyborg Buddha'). Consequently, it is also the case that, philosophically, many transhumanists ought to be classified as pessimists. I am among those who subscribe to this type of philosophical pessimism. Human lives are full of permanent suffering, including only short, temporary reliefs from suffering.

Such a description of transhumanism is very different from how transhumanists are usually presented in the mass media. Transhumanists are usually described as positive visionaries who think that in a few decades we will have conquered death, realized mind uploading, and metamorphosed into posthumans.[12] It needs to be realized that both descriptions can be appropriate ones. Actually, the philosophical pessimism is connected to the widely shared transhumanist wish to find a solution for dealing with suffering, and it needs to be realized that many technologies are responses for dealing with suffering. Positivity, in my words, stands for the attitude that technologies usually represent helpful responses for successfully dealing with the issue of suffering, which I explain in more detail in the following section.

1.2.2 Positivity

A naturalist account of the human implies that human beings have come about as a consequence of evolutionary developments (Sellars 1922). A concept of an unchanging human nature is rather implausible (Thomsen/Wamberg 2020). Even reason no longer counts as an entity which provides you with knowledge about the world. Reason has come about as a consequence of evolutionary mechanisms (Sorgner 2007, ch 2). Reason enables us to survive, to become more powerful, and to have some good times. Reason is not a device which generates truth for its own sake. Reason is not a unified entity which is identical in all of us due to its developmental origin. We have the organic predisposition to acquire reason, but in order to actualize reason, we need education. We do not know any words, and we cannot form sentences and make inferences, when we are born. Slowly, and gradually, our parents and our cultural surroundings provide us with these capacities (Sorgner 2018b, 2019c). We thus get upgraded. Education stands for various upgrades. We are steered by our educators. The *kybernaetes* is the helmsman of a ship, the one who steers and directs a ship. Our organism is directed by our educators. Hence, we are steered organisms, or in other words 'cyborgs'. We have always been 'cyborgs' since we became *Homo sapiens sapiens*.

Reason is a technology, a technology which has become a part of who we are. At the same time, it is a means which enables us to survive, plan, and communicate. It is a tool, but it has also become a part of who we are. We are who we are because of a technology having become a part of us; it is part of our hybridity. We are merely being forced by reason to see this analysis as a contradiction in terms. A technology cannot both be a means as well as a part of us. Why not? It can. Reason is not a capacity for providing us with knowledge of the world. It has merely helped us get to where we are right now, and it does not exist as a separate entity, as it has become part of our hybrid continual becoming.

We get upgraded with reason by means of our educators. Reason helps us to make plans, postpone and redirect our instantaneous desires, drives, and affects, and enables us to communicate with our peers about the issues relevant to us. There is no separate entity called reason. There is not one unified entity called reason. There are a great multiplicity of reasons. However, what we refer to as reason is sufficiently similar to what others regard as reason, so that communication usually works well. I order a coffee, and I get one. It is not that the meaning of coffee is the same for the bartender and for me. However, each of us had some kind of understanding of what the other person has in mind. Reason is a technology. It helps me to get a coffee in a bar. Hence, it is a means, a means which usually works quite well. Reason usually enables each one of us to realize the unfolding of our drives.

This is already one reason for having a positive disposition towards technologies. Reason is a technology which enables me to improve the quality of my life. In addition, technologies have become a part of who I am; reason, vaccines, the capacity to write, my ability to play the guitar … all of these abilities improve the quality of my life, which provides me with further reasons for being positively disposed towards technologies.

At the beginning of the 19th century, more than 90% of humans lived in absolute poverty. Even in the UK, more than 80% were living in absolute poverty. Nowadays the situation is different, and less than 10% of the world's population live in absolute poverty.[13] It is important to note that studies considered absolute poverty, and not relative poverty. This description does not imply that the world is perfect now and everything is going well. Every person who starves to death is a tragedy. However, this statistic is a reason for having a positive attitude towards technologies. Education, vaccination, antibiotics, and anaesthetics are technologies which have been responsible for increasing the average life span, and indirectly also for reducing the absolute poverty rate all over the world to 10% since the early 19th century.

Did the development of human flourishing not come with a certain cost? Did we humans not exploit and destroy the environment to get where we are right now? All acts come with a cost. As human beings, we all produce, cause, and are responsible for carbon dioxide emissions. These emissions cannot be avoided. If you wish to avoid them, you must stop breathing out. If you pursue this thought further, it might imply that it would be best if humans died out entirely. There are some critical posthumanists who demand exactly this. With lower carbon dioxide emissions, we would not have human-induced climate change, and we would have better living conditions. As there would not be any more humans, it would not make a difference to us whether living conditions were better or not, because living conditions need living entities, for whom the conditions need to be better. In whose interest would this be?

Instead of favouring the extinction of human beings it might be advisable to have just as many human beings on Earth as the world can sustain. However, in order to realize this thought, a world government which enforced eugenic practices would be needed, so as to guarantee that only a certain number of human beings were born. One would legally have to force humans not to have offspring. After our experiences with the German 'Third Reich'[14], this is not a social organization which we can regard as worthwhile. I strongly object to such an arrangement.

Negative freedom and plurality are wonderful achievements. However, they come with a cost. Humans strive for privileges. The less-privileged always attempt to overturn the system in order to reach a privileged position for themselves. Instead of aiming for a utopian ideal social organization, which cannot be achieved anyway, we need to find a different goal, an as-good-as-it-gets solution. Utopian goals have always led to decisions which sacrifice the present for a future which cannot be achieved (Sorgner 2018b). Utopias are merely in the interest of those who temporarily gain power, and enjoy the champagne with strawberries in a luxury villa next to the sea. Any utopia is dangerous, as it follows abstract goals without considering the implications on people.[15]

As-good-as-it-gets solutions should be what we aim for in our visions. If these are our goals, we are justified in taking a positive stance concerning our current situation, as, historically speaking, it is actually quite good for most human beings, despite the increased level of carbon dioxide emissions and the consequent great variety of implications concerning climate change. Climate change is a challenge, but I regard it as a challenge which we can deal with by means of new and innovative technologies. Instead of the demand to introduce new eugenic laws concerning procreation, or to get rid of human beings, or to return to a 'natural' world before the time during which evil technologies destroyed our harmonious relationship with nature, we desperately need to focus on technological solutions for the various issues which can be associated with climate change; for example, in vitro meat, roofs made from solar panels, real vegan cheese on the basis of gene-edited yeast, new architectural solutions for physical, biological, economic, and social conditions for successful and productive agriculture solutions in urban environments – for example Plantagon – and new means of transportation which are better for the environment, like Hyperloop.

There are good reasons for describing social, political, and cultural developments in a positive way. Besides the fact that the percentage of people living in absolute poverty has decreased from more than 90% to about 10% in the past 200 years, the average life span has increased significantly during the same period of time.

> [T]he health transition began at different times in different world regions; Oceania began to see increases in life expectancy around 1870, while Africa didn't begin to see increases until around 1920.
>
> Since then life expectancy doubled in all world regions.
>
> In Oceania life expectancy increased from 35 years before the health transition to 79 years in 2019.
>
> In Europe from 34 to 79 years.
>
> In the Americas from 35 to 77 years.
>
> In Asia from 27.5 to 73.6.
>
> And in Africa from 26 years to 63 years.
>
> Globally the life expectancy increased from an average of 29 to 73 years in 2019.[16]

Extreme poverty has decreased, and life expectancy has increased due to the further development and usage of technologies. Overall, this has significantly increased the probability for human beings of living a good life, which is a clear reason for holding a positive attitude toward technologies.

Besides mentioning abstract data, which show that technological innovations have improved the quality of life for most human beings, considering your daily routine and how much it is dependent on technologies should be sufficient for regarding a positive outlook as justified. You enter a warm bathroom in the morning, as the central heating works well. Clean water comes out of the shower. It is so clean that you can drink it. After the shower, you use the electric hair dryer, which reduces the likelihood of your catching a cold. You turn on the coffee machine and make yourself an espresso so as to have an energetic start to the new day. Then you open the fridge, which allows you to store fresh food. You eat a yoghurt for breakfast. While having your breakfast you can stay in contact with your friends, loved ones, and parents by having a video call with them, even if they live far away. Then, you clean your cups in a dishwasher, and your clothes in a washing machine.

A permanently increasing number of people have central heating, a hair dryer, a coffee machine, a fridge, a dish washer, a washing machine, and a mobile device which enables wireless video calls. Before 1900, none of these technological options was available. Nowadays, we can hardly imagine life without them; if you do not store an opened milk bottle in the fridge, you must throw it away after a few hours. These developments are widely available and taken for granted in developed countries. However, in less-developed countries too, more and more people can benefit from these innovations. Vaccinations and medicine for treating HIV-positive patients are being made available at a lower cost to more and more people in less-developed countries, too. In 2014, 41% of adults living with HIV received treatment.

In 2017, the percentage had nearly doubled, to 79%.[17] These few examples reveal the enormous impact technological innovations have had upon the quality of our lives. Many additional examples could be listed; for example, it was only in 1939 that the right to paid holiday leave became a legal right in the UK; the Grand Tour used to be a privilege which was reserved solely for the offspring of the gentry in the 18th and 19th centuries. These social changes are not unrelated to the processes of introducing new technologies, automation, and, nowadays, digitalization. Taking past developments into consideration can provide you with further reasons for having a positive attitude concerning the new implications of emerging technologies.

It is important to keep in mind what I said earlier concerning the concept of technologies. We often tend to think solely of smartphones, personal computers, and planes when mentioning technologies. However, our language use is already a technology. In the early 19th century, less than 15% of the world's population were literate. Nowadays, less than 15% of the world's population are illiterate.[18] Then, we had a world population of about 1 billion. Now, about 8 billion people live on the Earth.[19] This development is part of an increased usage of technologies, which has further social implications, as literacy is a capacity, a power, which has political implications. Hence, it can be stressed that increased literacy correlates with an expansion of democratic political structures. Since the beginning of this millennium, we have more democracies than autocracies in the world, and the gap between the two types of political regime has widened further since then.[20] The more educated people are, the more they demand that their voices be heard politically. This is part of the process which is characterized by an increased usage of technologies, as language is a technology, too.

The preceding reflections show the line of thought on which a positive attitude concerning technologies can be based. At the same time, you can hold the position that life is full of suffering, which is related to the continual becoming. As most transhumanists affirm a version of naturalism, this position is widely shared among them. This is the line of thought which leads to the judgement that transhumanists can be characterized as positive pessimists. In the next section, the relationship of positive pessimism with nihilism will be clarified further.

1.2.3 Nihilism

If you live in a world of continual becoming in all respects, nihilism follows. Two different types of nihilism play a particularly important role, that is, alethic nihilism and ethical nihilism (Sorgner 2010a). Alethic nihilism is an epistemological position, while ethical nihilism is a judgement on values. Both follow from the absence of ontological stability. If a judgement is true, it needs to correspond to facts. However, if there are no facts, but only continual becoming, then a judgement cannot correspond to anything. If a judgement corresponds to a Platonic form, then a truth in correspondence with the world is possible. Without the existence of such forms, in a world of continual becoming the best we can be confronted with are contingent nodal points. These, however, are important for our survival. A contingent nodal point is an interpretation. Whenever we make a judgement about the world, then we present interpretations. Each judgement is such an interpretation, which does not mean that each judgement is false. It merely means that each judgement can be false, which also applies to the epistemological position of perspectivism. It is one interpretation among many others. However, it is a plausible interpretation from my perspective, as so far no epistemological perspective has been presented which actually represents a truth about the world which corresponds with the world.

What about the judgement that I am writing on a PC (personal computer) right now? This is correct, of course. However, this is merely a pragmatic utterance. We know what PC means, and what we refer to when we say the word PC. However, what a PC is, is uncertain to us. Is it a continuous entity, or a discrete one? Does it possess mental aspects? Can there be carbon-based PCs? Hence, we have some idea of what the word PC refers to, but ontologically it is uncertain what a PC is.

What about the judgement that water is H_2O? This is, rather, a tautology concerning how we define hydrogen and oxygen. What hydrogen and oxygen is ontologically is uncertain. Judgements are either tautologies or pragmatic truths. Once we analyse judgements philosophically, they are interpretations. This judgement can be falsified only if it can be demonstrated that at least one judgement is not an interpretation but corresponds with the world. As this has not been achieved so far, it is plausible that all judgements are interpretations.

It is this epistemological stance which stands for alethic nihilism. Nothing positive can be said which corresponds with the world with certainty. In addition, alethic nihilism is not only a plausible epistemological stance but it is one which has ethical implications, as it goes along with a less violent stance concerning others.[21] If you are aware that your own perspective is an interpretation and you cannot establish its own ultimate superiority, you have to take a less violent stance concerning other people's point of view (Vattimo 1997, 40, 46–47, 66). They might make sense, after all. They might be appropriate for them. They might even be true. Alethic nihilism implies a less violent stance concerning others. Being less violent against other persons enables a greater plurality of human flourishing. One's own freedom ends where the freedom of the others begins. Once you harm another person, your acts become morally problematic. What exactly constitutes a harm is a debatable issue (Sorgner 2020a).

The stance that it is morally wrong to directly harm another person is no ethical knowledge either. We agree upon it, and this is why it is effective. It is a contingent nodal point to which we subscribe, and I am very happy about living in circumstances where this insight is widely being subscribed to. It is an achievement to live in such cultural circumstances. These circumstances are the result of a long history of power struggles during which people of various interest groups have been fighting for their right to live in accordance with their very own idiosyncratic wishes, desires, and fantasies, and there does not seem to be any drive which is so fundamental that it promotes the flourishing of all persons. This is what ethical nihilism stands for. It is the ethical perspective that no non-formal judgement of the good is plausible for all people. If you wish to make a judgement concerning an element which necessarily improves the quality of life of all people, you are bound to fail. Transhumanists take this insight seriously, which clearly comes out in the demand to treat the right of morphological freedom as a human right. Further rights which are widely shared among transhumanists, and which I uphold too, are educational freedom and reproductive freedom. Educational freedom includes people's parental right to genetically modify their offspring, as traditional parental education and genetic enhancement by modification are structurally analogous procedures. Reproductive freedom includes the right to select children for implantation after IVF and PGD, as selecting a partner for procreative purposes and selecting a fertilized egg for implantation purposes are structurally analogous procedures, too. Whatever is parallel also needs to be treated structurally analogously from a normative perspective (Sorgner 2016e).

Politically ethical nihilism emerges among transhumanists, as no serious transhumanist upholds a non-liberal position. There are a wide range of liberal and libertarian positions that have been affirmed among transhumanists. Max More represents the more libertarian end, whereas someone like James Hughes clearly affirms a more social-liberal democratic political system (Hughes 2014). However, no serious transhumanist is in favour of authoritarianism. Authoritarianism often is justified by reference to a non-nihilistic ethical perspective, for example, if you regard a certain concept of the good as universally valid and that good ought to be prioritized over right, then you have a reliable foundation for arguing for an authoritarian system.[22]

1.3 Conclusion

I have explained why transhumanism can be seen as a nihilistic, positive pessimism, and what the underlying lines of thought for affirming such an approach are. Unfortunately, most naturalists have not properly realized how close they are to postmodern intellectuals, and vice versa. A proper naturalist should embrace a postmodern alethic nihilism, which again can be explained most plausibly on the basis of a naturalist ontology of continual becoming. Given the permanence of becoming, philosophically one ought to talk about a pessimistic ontology, as it was affirmed by Schopenhauer, Nietzsche, or Buddhism. I often use the term naturalism, as this was the term used by many transhumanists to self-identify their own underlying world-view. Naturalism implies the affirmation of empirically accessible entities. Naturalism should not be identified with a naïve materialism. Materialism in its naïve version is an essentialism, which would undermine the alethic nihilism as I have specified it here. It would be self-defeating. A naïve materialist has to be caught in the Cretan liar's paradox, which does not apply to an enlightened transhumanism as I have presented the approach. In addition, naïve materialism implies that there is no mental realm but only a material one. Being in pain, having consciousness, and feeling emotions are experiences which cannot be identified solely with the term matter, as they undermine the characteristics associated with matter. When I talk about naturalism, it would be false to assume that I am assuming a naïve materialist understanding of the world. A philosophically informed naturalism might have more in common with a Spinozan ontology than with a naïve materialism. This is definitely the case for the psychophysiological ontology of continual becoming in all respects, which is central to my reflections. The suffering which goes along with continual becoming is the central ethical challenge with which we are confronted. Technologies are the best support for enabling us to deal with the suffering, as historically informed data analyses reveal. I explained this in the section on positivity.

Even though the current situation is not without challenges, it has never been better for people on Earth. Of course, there are quite a few tricky challenges with which we are confronted. Climate change is definitely one of the most crucial issues in this respect. However, the only plausible way of dealing with it is by considering the usage of emerging technologies. A return to a pre-technological state of nature, which never existed, is neither a realistic nor a desirable option for us. Despite all these challenges, there is no solid reason for not having a positive attitude. Our health spans are longer than ever before, the quality of our lives cannot even be matched by that of kings of the past, and even on a global level it needs to be stressed that we have managed to reduce the absolute poverty rate from over 90% to 10% since the early years of the 19th century, which implies a significant reduction of individual suffering.

Even though there is not a single element which is essential for a good life, which is what is implied by ethical nihilism, there are some factors which are shared by many people. The most important factor is that of the health span, as the majority of people identify a longer health span with a better quality of life. This does not mean that your life quality is being increased if your health span is increased. However, as this is the case for the majority of people, this is a factor which is politically relevant and which deserves much further attention than is the case so far. Being an ethical nihilist does not mean that one cannot have any standards. It merely means that essential standards no longer apply. All principles turn into contingent nodal points which lose their universal validity, and together with this the violence which goes along with such a validity claim. When Vattimo (1997, 46–47) talks about the violence which used to be associated with the sacred, this is what he means. It is a validity which knows no exceptions, which assumes to be valid, no matter what people claim themselves, and which deserves to be actualized against all odds, as this is the demand which goes along with such a validity. It is this attitude which is no longer plausible in a philosophically informed transhumanism, which can be identified with a nihilistic, positive pessimism, as I have described it in these reflections.

On a Silicon-based Transhumanism

In the Middle Ages, scholars discussed how many angels could dance on the head of a pin. Nowadays we talk about Bostrom's simulation argument, as it was popularized by Elon Musk. Both topics are fun. Both discourses make sense from the perspective of the specific cultural background. In the Middle Ages we had a Christian background. Now we cherish the sciences and technology. Yet, by dealing with these questions, we avoid being concerned with the most pressing issues of our times.

I progress as follows. Firstly, I will argue that there is no pragmatic need to be concerned with the simulation argument. Digitalization and automation are important developments. However, the most pressing moral issues concerning digitalization do not have to do with our being threatened by a super-intelligent AI. They have to do with total surveillance, privacy, and negative freedom. Secondly, I will show that we need to radically reinterpret the relevance of digital privacy, embrace total surveillance, and accept the collection of digital data for a democratic purpose. Thirdly, I will further reflect upon the following issue to provide some information on the question of how such a system could be structured: what could it mean to cherish negative freedom, to respect a person, and to avoid harming a person? Thereby, it will become clear that a democratic usage of our digital data is a pragmatic necessity. My philosophical ideas are intended to provide grounding for further reflections.

2.1 Transhumanism without mind uploading and immortality[1]

I have heard transhumanists claim that mind uploading is the crux concerning whether someone counts as a transhumanist or not. This is not the case. Julian Huxley, who first coined the term transhumanism in 1951, would not be a transhumanist if you had to believe in the possibility of mind uploading.

I do not regard mind uploading as impossible, and I definitely hold that we can and should use technologies to move beyond the current limitations of our existence. However, gene or cyborg technologies are far more likely possibilities of fulfilling this goal in the near future (Sorgner 2018b). Gene technologies cover the wide range of options from gene editing via gene analysis to selecting fertilized eggs after IVF and PGD (Sorgner 2016b, 140–189). Cyborg technologies have to do with the digitalization of the lifeworld, smart cities, the Internet of Things, and the upgrading of human beings by means of RFID chips. We have to realize that smart cities need upgraded humans and the Internet of Things needs to integrate the internet of bodily things. Both of these types of technologies are progressing rapidly, due to the central relevance they have for promoting a widely shared human goal, the prolonging of our health spans.

When it comes to transhumanism, what is discussed in newspapers, magazines, and the popular media most often is the possibility of mind uploading. It is an intriguing idea indeed. It also serves as an entertaining basis for Hollywood movies. The movie *Transcendence* by Wally Pfister with Johnny Depp in a leading role provides an excellent case in point. Not only does it provide non-experts with a helpful visualization of how mind uploading can be imagined but it also moves beyond fiction by including Elon Musk, who is among the best-known popularizers of this theory, in the movie. In one scene he sits in the lecture hall listening to a talk about mind uploading, referred to as transcendence in the movie.

2.1.1 Elon Musk and the simulation argument

In the non-fictional world Elon Musk regularly advertises and talks about the simulation argument in front of economic, political, and social leaders. This argument is closely related to the concept of mind uploading, as I will explain in more detail soon. The simulation argument goes back to an argument by Bostrom which he presented in his paper 'Are You Living in a Computer Simulation?' (Bostrom 2003). Musk adds some further twists which make his line of thought particularly imaginable.[2]

Musk reminds us that less than 50 years ago the computer game Pong was popular. Its rules are comparable to those of table tennis, and its graphic consists of lines and dots. The two rackets of the players and the centre line are represented by vertical lines, the ball by a point. These simple graphic elements made up the cutting-edge game of the 1970s. Meanwhile, Musk correctly points out that millions of players can now play simultaneously on the internet in a photorealistic environment. This leads him to raise the following questions: what will we be able to do in 2070, if the speed of innovation in this area is maintained? Will we then still be able to distinguish between artificial and fundamental reality? Musk is convinced that artificial and fundamental reality will be indistinguishable in terms of their experiential content. If this consideration is correct, then it is likely that we are already living in a computer simulation today, he concludes. But why should it be likely that we are already living in a computer simulation when it can be expected that in 50 years the experiential content of an artificial reality and the basic reality will be identical? There clearly seems to be a fallacy here, one might think. However, if simple probability calculations are included in this train of thought, the idea underlying the simulation hypothesis can be understood rationally. Here it is necessary to reflect on the numerical relationship between the future simulated reality and the currently experienced reality. This idea can be illustrated by means of a special example.

Musk's line of reasoning includes the assumption that lifelike computer simulations of the past will be possible in 50 years' time, due to the exponential growth in computer performance power. This idea presupposes that we will not be extinct by then and that some people will be so advanced that they are capable of lifelike computer simulations and, on top of that, that there will be an interest in computer simulations of the past. I currently regard myself as living in the year 2021 as one of about eight billion people on Earth. In future computer simulations of the past, the same will apply to every single inhabitant. Many people in every simulated world will expect to be one of eight billion people on Earth in 2021. Each of them will experience their own artificial world as intense, real, and emotional as each of us experiences the real world at present. From Musk's point of view, it is likely that there will be a billion such simulations of the past, resulting in the following consideration: why should I be one of the people who are in the one basic reality? The probability of this is low. In actual reality in 2021 there are eight billion people. In each of the simulations of 2021, of which there will be one billion (he assumes), another eight billion people will feel as if they are in the one fundamental reality. The experiential content of each person in each of the simulated worlds is in no way different from my own, since all kinds of experiences can be simulated in a perfect way, due to the enormous performance capacities of the computers. To calculate how plausible it is that I am in the basic reality depends solely on the mathematical probabilities. Am I one of the few who is in the basic reality or one of the many where this is not the case? If we assume one billion simulations, the probability would be one in a billion that I am not in a computer simulation. This consideration has convinced Elon Musk and other thinkers and entrepreneurs that it is almost certain that we are already in a computer simulation.

2.1.2 Implicit assumptions of the simulation argument

Is he right? Are his reflections compelling indeed, or at least plausible? What is crucial is that in the context mentioned it is impossible for the inhabitants of the computer simulation to be able to distinguish between the simulation and base reality. We are not there yet. By simply taking off our VR (virtual reality) glasses, we return to base reality. Even if we had a VR dress, we would simply have to undress to return to our everyday world. Musk has a different vision in mind. He thinks that it is highly likely that we are already within a computer simulation. However, if we are within a computer simulation and it is not so easy to drop out of it again, then it must be the case that we, too, have to consist of a binary code, ones and zeros on a hard drive. This has to be the case so that it is impossible for us to distinguish between simulated and base reality. This is also the reason why I have mentioned that mind uploading and the simulation argument are connected, as the argument provides reasons for claiming that we will be able to realize mind uploading soon. If we trust Kurzweil's predictions, it might be available in about 30 years' time (Kurzweil 2006).

In principle, this seems like a plausible argument. It receives further plausibility due to the fact that, since Darwin, fewer and fewer people claim that human beings have a categorically exceptional status in the world due to their immaterial soul (Sorgner 2010a). More and more people are embracing a philosophy of mind which implies that mental processes are the result of certain evolutionary processes, and that we need to continue developing further an embodied theory of mind. If such an approach is correct, then it would indeed be the case that there is no categorically ontological difference between mental and digital processes. Both types of processes consist of the fundamentally same type of something. Furthermore, it could be added that it is also clear that it is possible to transfer our mental content. All parts of our body get replaced every seven years. However, there remains a continuity of our mental content, of our memories, and our intellectual knowledge. This is a reason for maintaining that mental contents are not necessarily connected to a specific ontological something. Maybe this also supports the plausibility of upholding a functional theory of mind, whereby the mind can be compared to the software which runs upon a specific hardware, our body. It is this analysis of the mind which was centrally relevant for many transhumanists in defending the possibility of mind uploading (More 2013, 7).

It must be noted that a specific ontology has to be given for mind uploading to be a realistic option, namely one in which nothing must get lost during the transformation from an analogue to a digital entity. If our personality does not exist as digitally structured, it cannot be turned into a digital entity without any losses. What do our personalities consist of? Do they consist in a type of continuous being, or are they based on discrete entities? Both options are possible. If it were the case that energy can only occur as an integer multiple of the Planck constant, which is scientifically plausible, then the entire world could be a digital one. All entities could then be analysed as ones and zeros. However, the wave-particle dualism of quantum physics is an indication that other interpretative options are possible on a scientific basis. If a process can be interpreted only by means of a wave, then this process suggests a continuum rather than the possibility of finding digital entities.

Furthermore, what was not sufficiently considered in the arguments in favour of mind uploading was the distinction between living and non-living entities. Even though, post Darwin, many scholars no longer uphold a fundamental ontological difference between living and non-living entities, the quality of being alive is one which so far can be attributed only to carbon-based entities. We are not familiar with any silicon-based entity to which we can ascribe the quality of being alive, and computers are silicon-based entities.

There are two dominant ways for how to respond to this thought. Firstly, this thought can be questioned by referring to computer viruses. Should not computer viruses, which are self-replicating entities, be analysed as being alive? Secondly, it can be referred to genetically altered living tattoos, which have been realized by the Massachusetts Institute of Technology (MIT) in 2017.[3] Those can be seen as an initial step towards the development of organic computers. Here, it needs to be noted that with the option of a carbon-based computer we would indeed have an increased possibility of realizing mind uploading. We would 'simply' have to transfer the mental content of some organic stuff to some other organic stuff, which happens anyway, due to all human cells being replaced within a seven-year period. Even though such a transfer would still be amazing, it would not succeed in realizing the goal which many have associated it with and due to which mind uploading has gained so much attention in the first place. Mind uploading has often been identified with the possibility of radically improving our life span by transferring our personality from our delicate physiological bodies, which are easily prone to defects, to silicon-based hard drives, which are much more resistant. Hence, uploading our mind to a carbon-based computer would not significantly improve our situation in the world. Hence, we are left with the first line of thought: can silicon-based life be realized? Have we realized such life forms already? Should computer viruses count as living entities?

This is indeed a problematic question. An ancient analysis of what counts as a living entity has identified life with the capacity of self-movement. If this analysis is plausible, then we could indeed regard computer viruses as being alive. Stephen Hawking held this position.[4] However, even though they share many characteristics with living entities, they do not consist of cells and they do not consume organic matter, that is, they do not possess a metabolism. For the same reasons, other viruses do not count as being alive either. Yet, if we applied these criteria, we might simply refer to a tautology. Computer viruses are no organic organisms. This is what we have known all along. What we can say is that it is a good argument to refer to computer viruses when the claim is made that we do not know any living digital entities so far. Computer viruses are indeed an example which we need to consider seriously. It is at least not yet entirely clear that computer viruses ought to count as living entities. Yet, they represent the strongest reason for claiming that mind uploading will eventually be possible, because one would want to remain a living entity before and after the successful process of mind uploading. If life is impossible on a carbon basis, then we do not have to consider mind uploading any further, as then it would be impossible to transfer a human personality to a hard drive without any loss. It is what we would expect for a successful mind upload, as we would want to remain alive after an upload. Without being alive, it also seems impossible to imagine the coming about of consciousness, and self-consciousness, as we do not know any conscious or self-conscious entity which is not alive.

Consequently, if we do not regard computer viruses as living entities, we do not have a reason for upholding the possibility of mind uploading, as without knowing any silicon-based entity which is alive we would not have a solid reason for claiming that within the next 30 years we will create such entities and we will even be able to transfer our own personality onto a silicon base.

2.1.3 Immortality now

Mind uploading is often identified with the possibility of being able to make humans immortal. If we can upload our personalities to a computer, we can permanently move them onto different groundings so that they can stay alive for infinitely long. Rothblatt talks about a 'digital immortality' (Rothblatt 2015), and Kurzweil about the possibility to live 'well forever' in this context, but, unfortunately, I do not regard these visions as realistic ones. I appreciate both of them as thinkers as well as entrepreneurs. However, in this context we have to philosophically keep in mind what 'immortality' stands for. It can mean either that it is impossible for a living entity to cease to live or that it does not have to be the case for a living entity to cease to live. If we take any of these two literal meanings of the word 'immortality', I am bound to stress that neither of them can even be conceptualized.

Most transhumanists uphold a naturalist ontology, which implies that in principle all aspects of our lifeworld are empirically accessible. We have had one (or more) Big Bang(s), the universe has expanded, and eventually it will come to a complete standstill. Alternatively, it could also be the case that the process of expansion will become reversed, such that a contraction will occur and, in the end, a cosmological singularity will occur. No matter which one of these possible futures will occur, human beings, even as uploaded personalities, will not be able to survive any of these states. It is impossible to even conceptualize immortality if we start off with a naturalist ontology. Maybe it can be imagined that, together with a cosmological singularity, space and time will be dissolved, and we will survive in a radically different otherworld. However, such visions of immortality belong in the realms of fiction, religion, or psychiatry. They do not deserve to be considered any further here, as there are no reasons why any of this ought to occur.

Neither singularity nor immortality is near. Given a naturalist ontology, we cannot even conceptualize immortality. A thoughtful transhumanist ought not to use the term 'immortality' and mean it literally, at least not when she talks about realistic goals which can be achieved in the coming decades. As I said earlier, I cannot exclude the possibility of mind uploading. Yet the claim that we will soon have achieved digital immortality is not one which ought to be made by a critical thinker, as we have no reasons for regarding it as a realistic short-term option.

Luckily, most well-known transhumanists do stress that they use the word 'immortality' in a metaphorical manner. They use it in order to stress the relevance of increasing the human health span. It is important here to talk about the health span and not the life span, because we do not live just to survive. Life can be unbearable. If pain is too intense, there are people who do not wish to continue to be alive. If the will to survive were our fundamental drive, people would not commit suicide, or we would have to classify everyone who wished to commit suicide as mentally disturbed, which would be a violent and paternalistic act. The wish to commit suicide can be an authentic one. Consequently, many transhumanists prefer to talk about the relevance of an increased health span, which most people identify with, or at least a prerequisite of a flourishing life. If you are not healthy, your quality of life is usually reduced.

In this context, it is important to use relativizing phrases, which is an insight Aristotle put forward when he stressed that ethical insights are never necessary but usually apply for the most part or usually, if they apply at all. Hence, an increased health span usually promotes the quality of human life, be it as an intrinsic or as an instrumental good. By highlighting the relevance of immortality in a metaphorical manner, this insight gets put forward. It represents a way of stressing the importance of promoting an indefinite prolonging of the health span, about which I will say more in the next section.

2.1.4 Conclusion

As I said in the beginning, I have heard transhumanists claim that mind uploading is the crux concerning whether someone counts as a transhumanist or not. In this section I wanted to point out that – and why – this is not the case. In order to do so, I have analysed the plausibility of the simulation argument, which is central for stressing that we will live as uploaded minds in a few decades' time. Even though I cannot exclude the possibility of mind uploading, we currently have no reason to claim that we will eventually be able to realize it. The most important reason for not regarding mind uploading as highly unlikely is if you regard a computer virus as a living entity. Whether this is the case or not deserves further investigation. One of the main reasons why many transhumanists find the concept of mind uploading so appealing is their belief that it could enable us to realize immortality. This is not the case. We cannot even conceptualize immortality. Only if we mean immortality in a metaphorical sense is it plausible to use this word. It can be and often is used to highlight the relevance of increasing the human health span indefinitely. This is a widely shared human goal indeed. Most human beings identify an increased health span with a better quality of life. I agree. There are many possibilities for how this can be realized. We have significantly increased the human life span already. Since the early 19th century, it has doubled in Europe. Yet, some more radical approaches are currently suggested. Instead of treating a specific disease, one ought to reconceptualize ageing as a disease, and attempt to put a lot of effort into undoing ageing. An even greater potential for increasing our health spans can be identified with the option of transferring selected genes of other animals to humans. This is where transhumanism becomes interesting, as here we find realistic options for actually increasing the likelihood of humans living better lives. Gene technologies and upgrading humans by means of RFID chips are highly promising technologies for realizing human goals.

The simulation argument and mind uploading help Elon Musk and friends to get attention from mass media. However, when it comes to the pressing issues of our times, we ought to deal with other challenges. If we are concerned whether AIs will wipe us out, we avoid being concerned with the most pressing issues of our times. The genetically new human and the implanted new human are realistic options for moving beyond the current boundaries of our existence, towards a trans- and eventually even posthuman future.

2.2 A democratic usage of our digital data as pragmatic must-have?

Smartwatches, the Internet of Things, and a permanently increasing number of autonomous cars are nowadays an integral part of our lifeworld. It would be naïve to assume that herewith these developments have come to an end. The last common ancestor of human beings and great apes existed about six million years ago. The commercial use of the internet was established less than 30 years ago. We need to acknowledge that the digital age is still in its embryonic state, and it has already had significant effects upon our life-styles all around the world.

Digitalization processes also alter the potentials of other emergent technologies, among which the great variety of gene technologies is particularly noteworthy. The gene scissor CRISPR/Cas9 might even be the most important scientific and technological invention of the 2010s. Yet, without the application of digital technologies to genes, which is being done by means of Big Gene Data, gene technologies could not realize their full potential. The greatest potential for radically altering our way of life can be found at the intersection of these ground-breaking technologies (Stiftung Datenschutz 2017).

All processes of the lifeworld become digitized. Autonomous cars are taking over the streets. Blockchain technologies decentralize the internet. Cryptocurrencies attack the relevance of banks. Smart cities get developed. Yet, if humans remained the same, all of these processes could not unfold a significant part of their impact. Smart cities need upgraded humans. Elon Musk's Neuralink and all the other companies, institutes, and taskforces which work on brain–computer interfaces will have the most significant impact on the future of human flourishing within the coming decades (Sorgner 2017b).

Computers are in the process of getting smaller and of entering our bodies so that we turn into upgraded humans, who can interact efficiently with their environment in smart cities, and have the appropriate means for dealing with ageing, the worst mass murderer in the world. This development goes along with new challenges related to digitalization, whereby the coming about of the internet panopticon is the most serious of all of them. A social condition arises in which all of our digital traces could be connected with each other and could be under permanent surveillance. From 2014 onwards, China has embraced such digital possibilities and has associated a social value to each action.[5] Europe, on the other hand, has introduced rather rigid data protection regulations which undermine the possibilities of data collection and the realization of corresponding insights by means of establishing correlations, for example between genetic data and health issues. Both political attitudes have an enormous amount of political implications. From a naïve point of view, it could be said that China promotes security whereas Europe cherishes the norm of freedom, which is a central achievement of the Enlightenment process. However, such a simple-minded and simplistic dualistic analysis does not do justice to the multifaceted implications of the new uses of digital technologies. There are a great number of plausible personal as well as political reasons for digitally collecting data.

2.2.1 We have always been cyborgs[6]

After the shift of information processing from the analogue to the digital world of computers, the process of mobilizing these systems has emerged, from cumbersome PCs to much more practical smartphones. However, in order to be able to access digital information even more quickly and easily, and in order to guarantee that an efficient interaction between us and autonomous cars, the Internet of Things, and all other aspects of a smart city can be realized, it is necessary to integrate computers more intimately into our bodies. This is exactly what we are working on intensively. The individual components that currently exist in the smartphone will therefore have to be replaced by new devices. The computer monitor has become a smartphone interface, which is now in the process of connecting ever closer to the human optic nerves. Digital glasses, such as the Google Glasses which are no longer produced, are only a transitional medium in this respect, and will be increasingly integrated into people. Google has already developed contact lenses that are able to measure the glucose value of the eyes' tear fluid. Diabetics have to check glucose levels daily, which is usually done by taking blood samples. The implantation of lenses in the eyes and the subsequent immediate stimulation of the optic nerves would be the obvious next steps in this development. If there is no user interface left, it makes sense to also integrate the chip into the body. At present, the outside of the hand, between the thumb and index finger, is widely used for having a chip, as it can easily be used to open doors. A Swedish company already offers its employees such a chip for this purpose. It is obvious that this system will be used as a versatile key replacement in the not-too-distant future, for example for hotel rooms or for opening and starting cars. However, the possibilities of such a chip go far beyond this simple use, since it is in principle already today an equivalent computer replacement. Without an additional external device, the control of such a chip must also be revolutionized. Just as the mouse and keyboard have been replaced by swiping and speech technology, new control elements are also required for an implanted computer. As early as 2016 a student of mine gave an in-class presentation using the control wristband Myo, which is connected to the computer via Bluetooth and enables the wearer to control a Power Point presentation by means of gestures. Gesture control systems radically change the choreography of man–machine interaction. A system integrated into the body could also be operated in this way, so that swiping over device surfaces and an external mouse are no longer required. Neither would an analogue or a digital keyboard be necessary for text input. Voice commands can already be employed. Meanwhile, however, intensive work is being done to avoid such commands by trying to translate thoughts directly into digital information. This means that only thoughts would be necessary to compile a text by means of a brain–computer interface. A team led by Tanja Schultz from the University of Bremen was responsible for fundamental research in this field.[7] In the meantime, Facebook has taken up this idea and employs a team of 60 people to put this knowledge into practice[8]. The future of typing is thinking. The personal computer turned into the smartphone, which will become reduced to the size of a small chip by means of which our bodies will be upgraded toward their transhuman existence.

Is this a categorically new development? Are we going beyond our previous humanity here? Are we becoming cyborgs now? It is central to the assessment of these new technologies that we have always been cyborgs (cf Haraway 1991; Hayles 1999). The word 'cyborg' means cybernetic organism, whereby cybernetic, from the ancient Greek 'kybernetaes' (κυβερνήτης), means helmsman. Cyborgs are therefore controlled organisms. Control already happens with us becoming human. In philosophy, human beings were usually defined by their ability to speak. Learning language is our first upgrade, which our parents provide us with. Our cyborgization continues with the acquisition of new skills, such as learning mathematics, history, and so on. However, a new dynamic is currently emerging. Control is being exponentiated, for example through genome editing (gene modification) and brain–computer interfaces. Ever smaller chips are migrating into our bodies, where they form interfaces with nerve cells or organs in order to collect valuable information about our bodies, using sensors. These technical developments continue a process that began with our first upgrade in language development. With the integration of digital technologies into our humanity, new possibilities as well as serious challenges arise. Both aspects are related to total surveillance.

2.2.2 Personal interests in data collection[9]

This process has enormous advantages in many respects. For example, the constant monitoring of one's own body could be decisive for readiness to combat ageing-related processes (see Sorgner 2019c). As soon as the blood sugar level, cholesterol level or blood pressure seems to change in a problematic way, people could be digitally warned, so that the problem can be addressed as soon as it arises, and not only when it is well advanced. Even a predictive maintenance of humans might be possible on the basis of these technologies. Predictive maintenance is already being used in machines. Sensors in an aeroplane can tell us that a specific part of the engine is likely to become dysfunctional within the next six months. We can replace this part, so that no risk to human life will occur in this respect. With RFID chips entering the human body, we can realize the predictive maintenance of human health. Researchers of Tufts University have already developed a tooth-mounted sensor which tracks every bite someone is makes.[10] Further such sensors could make up an entire Internet of Bodily Things which could then interact with the regular Internet of Things. The possibilities associated with this type of body monitoring are enormous and are likely to be of significant relevance in combating ageing-related processes. Here, the aspect of human flourishing comes in. Technologies have always increased the likelihood of our flourishing, and human beings have a great variety of associated goals. Yet there are some challenges which are troublesome for most of us, and one of these challenges is the process of ageing (cf Ehni 2018). Two-thirds of all deaths are related to ageing processes.

The more personalized data we have concerning the correlations between genes and ageing, life-style choices and well-being, and genes and well-being, the more reliable our data is for biotechnological research and medical interventions. Hence, it is strongly in our personal interest that our personalized data should enter a big data computer so that correlations between genes and quality of life can be realized. Based on this research, diseases can be treated, ageing can be dealt with and human flourishing can be promoted. All these issues are in the interest of most human beings.

2.2.3 Ageing as disease[11]

We reach our best bodily performance capacities at the beginning of our 20s. From this point onwards, our body is continuously in decline. We age. Even our brain, whose development process is completed at the time of our highest performance, slightly loses weight from then on. Our capacity of sight fades. Our hair becomes thinner and grey. The skin's elasticity decreases, which is why wrinkles appear. Even if we continue to eat normally, there is an increase in weight. Mobility, stamina, and speed decrease. In men, the testosterone level decreases from the age of 20. In women, the decrease of libido begins a bit later. Even bone density is at its peak around the age of 20. These are quite common processes. We age. However, these processes are accompanied by damage that can directly affect our ability to survive. Nevertheless, they are widely considered as undoable ageing processes, and not disease states.

At the same time, we must also note that two-thirds of all deaths are linked to ageing-related conditions. Approximately 150,000 people die every day worldwide. Of these, 100,000 die from diseases caused by age-related damage. Only a small percentage die due to conditions caused by HIV/AIDS. These findings are linked to the assessment that ageing is a disease. The following processes are generally associated with ageing (De Grey/Rae 2010): (1) mutations in the genes, (2) mutations of the mitochondria, (3) deposits in the cells, (4) deposits outside of cells, (5) cell loss, (6) loss of the ability of cells to divide, (7) increase in extracellular protein cross-linking, which reduces elasticity between cells. The first process can lead to cancer, the fourth to Alzheimer's disease, and the fifth to Parkinson's disease. Nevertheless, it is primarily diseases that are treated, rather than the damage that leads to them, and which are identified with the general ageing process. This is the crucial problem. The limit of our average life expectancy could be constantly pushed back if the challenge of ageing were adequately addressed. It is this issue which could significantly promote human flourishing in all parts of the world (De Grey/Rae 2010).

We are doing quite well already. The fact that we are on average older than our ancestors is a fantastic development, and it is one that is linked to our being technologically developed, that is, by means of education, the acquired ability to read and write, hygiene, vaccinations, anaesthetics, antibiotics. All these phenomena are technical. They have all helped to double the average life expectancy in Europe since the mid-19th century; but also on a global level, the average life expectancy has increased significantly. This applies in particular to those countries where the implementation of these technical measures has been supported by the government.

Some researchers may argue that the absolute age limits of the human species have not changed, only that many more people are now able to come close to these limits, especially because of the significantly lower infant mortality rate. Even if this were the case, it does not mean that an upper limit for the human being is a necessary characteristic. Homo sapiens sapiens have been around for about 40000 years. Two hundred thousand years ago, *Homo sapiens* first came into existence[12]; six million years ago, the apes living today and *Homo sapiens sapiens* had common ancestors[13]. It would be naïve to assume that *Homo sapiens sapiens* will still exist in six million years' time. Species must constantly adapt to a changing environment. Either a species adapts, or it dies out. It is therefore necessary to constantly resort to new techniques, and first to develop them.

One promising research area in the fight against ageing is the creation of human–animal hybrids. Since 2008, research on human–animal hybrids has been approved in the UK. However, the hybrids must be allowed develop for no more than 14 days.[14] In 2017, a human–pig chimera was created at the Salk Institute for Biological Studies in California, the development of which was not interrupted until 28 days after fertilization.[15] Japan ruled in 2019 that it is lawful to give birth to a human–pig chimera.[16] Not only is it possible to create organs in this way, where the probability of rejection by the recipient is low, but also the possibility of transferring and integrating non-human genes into humans could be realized by means of this research. Numerous characteristics of non-human animals could well be in humans' interest when it comes to ageing. The axolotl genome is particularly remarkable in this respect. The genome of this salamander is about ten times larger than that of humans. After losing one part of its body, the axolotl grows a perfect replacement with bones, muscles, and nerves within weeks. Even if the retinal tissue is damaged and the spinal cord is severed, they can be restored. The jellyfish *Turritopsis dohrnii* has another amazing feature. This is the first multicellular organism known to us which can transform itself from the stage of a sexually mature individual back into a sexually immature form of life. Only after reproduction do these medusae die. Some specimens of the Quahog mussel are said to have lived for over 400 years. But also larger animals sometimes reach an advanced age. Greenland whales can survive for more than 200 years. The giant turtle Harriet, who died in 2006, had reached the age of 176 years. If we were able to successfully and reliably transfer the genes responsible for these age ranges to humans by means of the newly developed gene scissors CRISPR/Cas9, then many people would certainly be extremely interested in this option.

Another existing technique could also help to increase life expectancy, namely the application of big data analysis to the genes found in centennials. The company 23andme which was founded by the ex-wife of a Google founder, is extremely promising in this respect. Estonia offers free DNA testing to its citizens if they share the results with the government.[17] Kuwait implemented some ground-breaking regulations too, as it was made obligatory for all visitors and all residents to deliver their tissue samples.[18] If this was not done, a person could face a prison sentence of up to seven years. Without reflecting morally on these regulations, it is clear that with regard to the application of big data analysis to genes, these developments must be taken into consideration. On the basis of such research we might be able to discover that certain genes or a certain gene constellation are usually present in centennials, but that they are less common in people who die at an earlier age. Such a finding could indicate that there is a correlation between the corresponding genes and longevity. Gene modification or gene selection could then be used to promote the presence of the corresponding genes.

Here, we can see that, by means of a successful interplay between gene and digital technologies, the chances increase that we will be able to successfully fight ageing, that is, the processes which bring about damage. What we are doing so far, is investing a lot of money in dealing with cancer, Alzheimer or Parkinson. All of these diseases are often the long-term consequences of the aforementioned damage. It is politically imprudent to act in this way. If we successfully dealt with a specific type of cancer, the life span of an individual patient might increase by five years. However, if we successfully dealt with ageing, there would be an increased probability of gaining an additional 50 years, which would make an enormous difference in terms of the option of human flourishing on an individual level. Hence, the potential of applying digital technologies to the disease of 'ageing' is enormous. An increased political interest in directly supporting research concerning undoing of ageing, rather than focusing on the tip of the iceberg – that is, the diseases which are the long-term consequences of ageing-related processes – could promote human flourishing much more efficiently than is currently the case.

A further development, which was presented by MIT researchers at the end of 2017, could supplement such developments. MIT researchers have identified a technique that enables them to print three-dimensional living tattoos from genetically programmed cells. These can be used as sensors for the body's own processes, but also for environmental influences, so that we can always react immediately to environmental and bodily dangers. This development also represents the first step towards living computers, with the help of which it would also be possible to make use of the large number of previous advantages of the upgraded person.[19] We develop autonomous cars, smart cities, and the Internet of Things. In order for us as humans to interact efficiently with this technically improved environment, we too must upgrade to reach this new level of our existence as cyborgs. Smart cities and upgraded humans are two processes which need to be developed hand in hand. Even though these developments could be of enormous relevance for increasing our health span, which would have a significant effect in terms of the likelihood of human flourishing, new challenges arise which should not be neglected either. The internet panopticon has to be regarded as the most serious challenge which is connected to the digital age.

2.2.4 Political interests in data collection

The greatest potential for a radical change in our way of life lies at the intersection of digital and genetic technologies. A central prerequisite for the application of the latest techniques is the availability of data on correlations between genes and diseases and psychological and physiological characteristics, as these are the prerequisites for the application of improving genetic techniques, such as genome editing, selection after artificial insemination and pre-implantation diagnostics, as well as bioprinting (Sorgner 2016e).

The relevance of collecting a wide range of personalized digital data goes beyond reasons related to individual flourishing. Collecting digital data is also relevant for policy making, for international as well as national political decisions, as well as for all areas of economic processes. The following examples provide only a brief selection of reasons why not collecting data cannot be a realistic option for us.

So far, I have not yet mentioned the most important reason for accepting, entering further into, and embracing the internet panopticon. The central reason is that we live in a globalized world, and that data is the new oil, as many experts stress. This needs to be relativized. Oil is a natural product, while digital data is an intellectual property. Yet the implications can be compared, for data as well as oil means power as well as financial flourishing (Bacon 1859). Given this realization, not collecting digital data cannot be a realistic option. Digital data is a central pillar for economic flourishing. There are other pillars, like engineering or natural resources; still, the use of all the other sectors can only be realized in the best possible manner if data is taken into consideration too, and the importance of data will continue to increase in the future as more precise and specialized data becomes available. Countries and institutions which embrace the internet panopticon have the best possible foundation for realizing economic flourishing; for example, Google, Facebook, and China are the significant players regarding the collection of digital data. In the US, digital data is primarily being collected by big companies. The Chinese solution seems to be even more efficient, as in China a social credit system has been gradually introduced from 2014 onwards, which enables the collection of all digital traces. The amount of digital data which gets collected in this manner can hardly be overestimated. The more digital data becomes available, the more political, scientific, and financial well-being can be realized. Europe, on the other hand, has institutionalized data protection regulations which go against the possibility of the helpful collection of digital data. It had good intentions, but the regulations undermine many of the most fundamental European interests; digital data for scientific research, political decision-making processes as well as economic flourishing will not be available. And yet, it will be needed. It can be expected that the consequences for Europe will be devastating, as Europe will have to pay China to get hold of the data needed for all these enterprises. China will continue to collect more and more digital data and consequently will gain more economic as well as political power. Europe's economy, on the other hand, will not be able to compete with China's. The main reason why the Chinese will continue to visit Europe will be its rich cultural history and its great variety of fascinating culinary experiences, such as the Colosseum, the Louvre,

Neuschwanstein, pizza, Kaiserschmarrn, and the Swiss cheese fondue. We will get paid for performing traditional dances for the Chinese, and for dressing up in our local traditional clothing, for example as Roman gladiators. Yet there will be no significant interest in us as partners in research or for economic reasons. These implications of our not collecting data will have an enormous relevance for our financial well-being. It will decline significantly. The middle class will be the first interest group which will strongly realize the consequences of these developments. They will not be able to afford the latest cars, to have several vacations a year, or to work in well-paid jobs with a decent university education. Whenever people are worse off than they used to be, even relatively, they look for scapegoats; someone else must be responsible for their social state. Blame is usually afforded to members of minority groups, immigrants, or the weaker others in general. If this is the case, extremist parties will be elected, and tensions within a society will increase. Civil war becomes inevitable.

All these reflections make embracing the internet panopticon practically inevitable, as we cherish our personal as well as our economic well-being and do not wish to be worse off than our ancestors. Both interests mentioned earlier are closely linked to and can be promoted immensely by means of data collection. This is the central reason why the collection of digital data is a pragmatic must-have, and so far, I have not heard a good response that renders these reflections implausible or invalid.

I am not pleased about these reflections myself. Yet there seems to be no other way out. I wish to stress very much that my analysis is not one about which I am happy, as I am aware that the risks and dangers for a liberal system are enormous. Any structure of total surveillance could be used against the plurality of human flourishing and in the interest of those who have access to the data. This issue can become particularly critical if a political system has a totalitarian structure. There is the risk of the coming about of a totalitarian surveillance society of unprecedented scale, which would be devastating for the great plurality of human flourishing. However, even in a liberal system such surveillance structures can have extremely problematic implications. There is always an enormous risk connected to those who have access to the data and who could use the information in their own interests. This is the central challenge with which we are confronted. However, at the same time it needs to be kept in mind that we do not wish to give up the possibilities connected with big data analysis, either. Hence, I need to emphatically point out: we need to embrace the internet panopticon, which goes along with a loss of privacy, as an as-good-as-it-gets solution. Is this problematic? Does it have to have devastating consequences? Or is there a possibility of affirming plurality while still being able to use the benefits which go along with big data analyses?

2.2.5 Internet panopticon[20]

What is this internet panopticon, and how does it relate to the digitalization phenomena described here? The panopticon is first and foremost an architectural structure that can be used meaningfully in hospitals, schools, and factories. It was developed in the late 18th century by the founder of utilitarianism, the philosopher Jeremy Bentham, and was part of his proposals for legal and social reform. His main focus in this regard was the architecture of prisons. This was not realized during his time; however, prisons and factories were later built on the basis of his designs.

The panopticon has the advantage of permanent and cost-effective monitoring of prisoners. It consists of a circular building with numerous cells completely flooded with light, with windows on the inside and outside of the circle. In the centre of this circle is a dark tower with a few peepholes, in which the guards are located. The prisoners cannot observe the guards, whether they are inside the tower or not. However, the prisoners are aware that all their actions can be observed at any time. Compared to a conventional prison, the panopticon requires fewer personnel, which reduces operating costs without affecting efficiency. Prison inmates internalize their awareness of constant surveillance, with the result that they themselves are inclined to censor their own actions (see Foucault 1977, 202).

In prison, surveillance plays a role between guards and prisoners. In the internet panopticon – and any person who is living in a technologically advanced society is part of it – the situation is even more complex because the individual roles of guards and prisoners are less clearly defined. Any excellent hacker could be a guard. At the same time, however, every internet user is a prisoner. Digitalization processes and the Internet of Things improve our quality of life, but at the same time both enable total continuous monitoring. The most important areas of personal surveillance are: (1) our space-time location (GPS (global positioning system), public surveillance cameras, navigation systems); (2) our psychology (scanned content of e-mails, internet search queries, websites accessed); (3) our physiology (health data, biometric data, genetic analysis, see Liu 2012). Information from the first two areas is easy to access. At present, the data in the third category are particularly in demand; especially, the level of interest in genetic data is currently enormous and many companies, such as 23andme, offer services in this respect. With humans becoming more upgraded, all of these data will come together and be digitally available and accessible.

In the 17th century Francis Bacon already recognized the following: 'Nam et ipsa scientia potestas est' (Bacon 1859). Information is power. Abolishing or reversing digitalization is not an option. This is neither possible nor desirable.

However, totalitarian mass surveillance is not a cultural environment that should be considered desirable at first glance. The central challenge here is how to structure the cultural environment so that advancing digitalization does not go hand in hand with a retotalization of society on an unprecedented scale. The destruction of the internet is not a suitable solution for our situation. We want to preserve the internet and expand it even further. It promotes our quality of life. With the existence of the internet, however, we remain trapped in the prison of the internet panopticon. This also means a loss of privacy.

What is the challenge of privacy? Is it really that important to us? Why do we appreciate it? The two most important reasons in this regard are the theory of property and the theory of sanctions for privacy (Floridi 2014, 116). However, these are not mutually exclusive theories. Both theories are united by the recognition of the importance of power. This needs to be explained briefly. My privacy is important to me because it goes hand in hand with information about me. Information about me is my property. It is my intellectual property. Property is important again, because property is accompanied by power. She who possesses something may dispose of the thing possessed. This is power. The sanction theory of privacy can also be explained in this way. We value privacy because we fear being sanctioned for the information associated with private data; for example, when we privately do something that is either legally, institutionally, or morally reprehensible, such as consuming strong drugs, supporting euthanasia, living and loving in a polyamorous way. Sanctions mean restrictions, which in turn implies a loss of power, for example if someone is punished with imprisonment, exclusion from an institution, or social exclusion. As far as the first theory is concerned, in a system without privacy, at least nobody is better off in advance, except those who are guardians themselves. Since the role of the guardians is not clearly defined in the internet panopticon, however, here no one is not in the role of a prisoner. The guard can also be monitored. As far as the sanction theory is concerned, the situation is different. The more values and legal and institutional standards there are, the more sanctions each of us must fear. This is the decisive point to be addressed. How many sanctions should we have to fear?

Staying in the internet panopticon seems to imply the constant fear of sanctions. But is it possible to live without fear of sanctions? Hobbes has made it clear to us that this is not the case. Even the strongest must sleep once, and when you sleep, then you can be killed by the weakest. In a world without political structures, we must fear being killed whenever we are weak, asleep, or sick. This situation is in nobody's interest. This is why political structures are necessary. However, these mean that there are standards, which in turn make sense only if one is sanctioned in some form if one breaks them. Fear of sanctions comes with every meaningful political order. However, sanctions are imposed only if a breach of the rules becomes apparent, which in turn becomes more likely the more monitored a system is. On the other hand, this means that in a total surveillance state there is a very high probability that every rule violation will come to light. Does such a political order have to be problematic? Not necessarily. If the murderer of an innocent toddler, a rapist, or a kidnapper is caught, it is in the interest of the population. One problem, however, is that in no social order are the rules without flaws; that is, in any system there is a risk of sanctions for actions that are not necessarily morally problematic. Homosexual acts were prosecuted in Germany until 1968. Now, on the contrary, in Germany marriage is lawful for everyone. However, marriage is still limited to two persons. In Colombia the union between three men has already been legally recognized[21]. Incest is still forbidden among adults today, but this is not the case in Spain. Why a state should have the right to tell competent adults who they can have sex with is not clear to me. A sexual relationship is exclusively a contractual relationship between two (or more) competent adults.

One reason why the right to privacy is important to us is that there is no system of rules without its shortcomings, and we fear being sanctioned without good reason. For this very reason, however, the recognition of the norm of negative freedom, that is, the absence of coercion, is such a central one in law. The more individual possibilities there are, the less likely it is that one will be sanctioned for inappropriate reasons. In a system in which every form of privacy has been lost, this insight is all the more important. In negative terms, this means that the political norm of freedom is particularly important in a system of total surveillance, as it reduces the likelihood of inappropriate sanctions. The same point can and should be formulated positively, and this justification is at least as important as the previous one. The political norm of freedom is particularly important in a system of total surveillance, as it also increases the probability of enabling a great diversity of good lives. Here I assume, as I have often explained, that a non-formal concept of good living is highly implausible, which roughly means that no statement as to what constitutes a good life is necessarily plausible (Sorgner 2016b). Every person has a unique concept of what enables her to lead a good life. These considerations make it clear that the only convincing solution to the question of the best possible cultural framework against the background of digitalization can be that we must always strive to promote the norm of negative freedom at the legal, social, and cultural levels. However, the fact that negative freedom is important is not based on an insight. Rather, it is an individual judgement on my part that, fortunately, today is widely shared. Norms and values are just as much fictions as money is. In all these cases it is a matter of created, fictitious, and imaginary evaluations, which can be effective only if they are shared. In all my writings I advertise the efficacy of this norm of freedom. If this moral, social, and legal norm can be promoted, we have reasons to hold that we can benefit from the consequences of living as upgraded humans in a smart city, while at the same time we will be in the internet panopticon. Privacy is abandoned, but human plurality needs to be promoted. Then, total surveillance and the greatest possible plurality of human flourishing can occur hand in hand, our acceleration will lead to a further deceleration, and the average human health span will be significantly expanded, which further promotes the likelihood of human flourishing.

2.2.6 A European social credit system as a pragmatic necessity

These reflections show that there does not have to be a conflict between total surveillance and freedom, and there are strong pragmatic reasons why we need to embrace and accept total surveillance, that is, the loss of privacy. The most challenging issue is who has access to all the available personalized date which gets collected. A challenging pragmatic issue is that this data must be protected well so that hacking it is not a practically realistic option. My suggestion was that the data ought to be processed primarily by algorithms, so that the quantity of humans having the right to access this data is limited. In the end, this issue is a pragmatic one. My central claim in these reflections is that we urgently need to rethink the meaning of data collection. Instead of data being collected primarily for marketing purposes, as is being done in the US, or for realizing rigid political goals, as is done in China, we need to develop a European way to use data democratically.

Why should the data not be collected by private organizations? Are private institutions not better at problem solving? Are private companies not the ones who embrace and drive innovation? The central reason why data must not be collected by private structures is that data is power. The more data an institution possesses, the more powerful, and also politically influential, it becomes. If a private company had the right to gather all the data, it would be become a strong quasi-political player. In addition, it needs to be considered that all different kinds of data need to be collected. To guarantee that this can be the case, political decisions need to be taken so that the possibility of collecting all the data becomes an option. This could not be realized privately.

A central step for realizing the collection of all these data would be for us to become upgraded humans, for example for chips to find their way into our bodies. Chipping humans has to become the norm. In Sweden, several thousands of citizens are getting microchipped already every year[22]. However, offering the possibility of receiving a chip and making it politically obligatory seem to be two radically different procedures. Making it obligatory to chip humans seems to imply that the government forces a problematic harm on all citizens and visitors, which again seems to undermine the norm of freedom. This does not have to be the case. Let us compare this procedure with that of obligatory vaccinations. In both cases, we have a legal obligation whereby the bodily integrity of a person is challenged. However, in both cases it can be a legal obligation only if the risks associated with the procedure are minimal and the social benefits significant. This is the case. At the moment, this seems to be a radical step. Once the procedure has become the norm, it can be expected that it will be of no significant concern to us.

So, let us return to my claim that we need to develop a European way to use data democratically. The financial gain associated with the data collected needs to be in the interest of the people. In the US, it is in the interest of companies. If private companies have all these digital data, they turn into significant political players. Here, the danger arises that the foundations of liberal democratic structures will be undermined. In China, digital data is being collected by the state. However, the values and norms on which their system is based are incommensurable with the achievements of the Enlightenment. In Europe, we have currently rejected the usage of our digital data by having implemented rigid data protection laws. This is not in our interest, due to the possibilities related to using these data. We need to develop a social-democratic alternative which takes into consideration that the relevance of privacy is related to the fear of sanctions, as well as to intellectual property.

The fear of sanctions needs to be reduced in order to guarantee the affirmation of a great plurality of concepts of the good life. Hence, we need to embrace a radical plurality of concepts of the good. Sanctions should occur only when harm is being done to a person. This is the reason why we will reflect upon what it means to be a person as well as to harm a person in Chapter 4. Moral, institutional, as well as legal sanctions should occur only where proper harm to a person is being done. This issue needs to be clarified further.

If a government stores all digital data, and uses them, then it can be argued that expropriation has occurred, which would be an illegitimate harm being done to persons. However, this needs to be rethought. It would not be an expropriation of our digital data, that is, our intellectual property, if the data were used in a democratic way and were used so that it helped to finance our interests. Here, the issue of health comes in. The majority of citizens identify an increased health span with a higher quality of life. This matters politically. This is the reason why universal public health insurance is politically justified. Yet the costs of upholding such a system are enormous. Even in Europe, differences in the quality of universal public healthcare systems are enormous. Health-care is incredibly expensive. Yet, it is in our interest. If the digital data were used to at least partially cover the costs of universal public health insurance, it would not be an expropriation but, rather, the payment for a service which is widely requested. As having a health insurance is a widely shared human interest, it is a duty of the government to provide people with it.

Developing a new drug is risky and costs a lot of money. If a pharmaceutical company has successfully developed a new drug and has patented its invention, it has the exclusive right for 20 years to realize a financial gain from its patent. It can charge whatever amount of money it wants for the drug developed. This makes sense, as the company took the risk and financial burden to develop the drug in the first place. However, data is needed for developing new drugs. Where do they get the data from? In a political regime with a total digital surveillance system, the government stores and projects the available personalized data, and can pass depersonalized versions of the data on to others, for example a drug company. In this way, certain limitations can be imposed on the developing company. The pharmaceutical company can no longer charge whatever is in its interest, as the drug was developed on the basis of data provided by people. The data was made accessible on the basis of a contract with the government which limits the rights of the drug company. In this way, it can be guaranteed that newly developed drugs can be made available to the people on a financially more accessible basis, or in a way that they can be included in the universal health insurance. Hence, the storing and using of data by the government is not an expropriation but a payment. We support the payment of a universal health insurance by means of our personalized data. This is what I mean by democratizing the use of collecting data.

Nevertheless, it can be objected that even though I continually stress the relevance of the norm of negative freedom, and the need to further promote plurality, does not my claim that we all need to embrace total surveillance undermine the relevance of freedom? It is clear that it does. No society can have absolute freedom. Sanctions for certain behaviour are necessary. If someone kills an innocent person, the murderer needs to be punished. Can a certain type of bodily harm become legally obligatory? Vaccinations are the best example. Still, one can wonder whether it would not be more in tune with the norm of freedom, if opting-out were provided, that is, if it were possible not to be forced to pay for the universal health insurance by means of permanent surveillance. If citizens preferred to pay for the universal health insurance with money rather than with personalized data, should a social-liberal democratic society not offer its citizens this option? Negative freedom is a wonderful achievement. Maybe, this option should be available. However, what would be the consequences? How much would it cost to pay for the universal health insurance with money rather than with personalized data? If it did not cost much, then many citizens might choose this option, which would undermine the goal of collecting the data in the first place, and opting out would become more expensive. In this case, only very few rich citizens could afford it. Does this not undermine freedom, too? If the power difference between the rich and the poor becomes too big, then the weaker ones are under illegitimate pressure. Hence, freedom undermines itself if one provides citizens with the possibility to opt out. Even though it initially seems to be more in tune with negative freedom, if an opt-out were legally provided, given further reflections, it seems more likely that in this case freedom would undermine itself. In any case, further reflections and practical evidence are needed for further judgements on this issue.

A further and widely shared concern needs to be a mentioned, namely that of discrimination. Does not a digital system of total surveillance promote discriminatory structures, and maybe amplify them even further? The contrary is the case. The aforementioned worry applies only if digitalization is promoted and a person becomes suspicious on the basis of human-defined categories. It can actually be shown how full monitoring leads to a reduction in the violence caused by racial discrimination. People of colour often share the experience in the US and Europe, where white customers look at them with suspicion in the supermarket. It seems that they are being checked to see if they are stealing goods. One woman of colour reported her shopping experiences in that Amazon Go markets that were recently established in some US cities.[23] How she was met was different from in traditional supermarkets. She was not viewed with suspicion. In these markets, you have to register when you enter, which means that from then on you are under permanent digital surveillance. Because every single item in the store has an RFID chip, it is linked to your Amazon account when you take it and put it in your own basket. Once you have completed your purchases, you simply have to go, and the total cost of all the products you have put in your basket will be debited from your bank account. The aforementioned woman of colour relates that no one looked at her with suspicion when she was shopping in an Amazon Go market. It is clear that once you enter such a store, you will be registered and you will pay for what you take with you. So her shopping experience was not discriminatory. If such a system were generally established, it could even be expected that such changed behaviours would become common public practice. By changing a public practice, you also change widespread stereotypes. This example is an indication that we have reasons to argue that certain types of racial discrimination can be resolved by total digital surveillance. Many further specific issues could be discussed, but it was not my intention to develop a fully worked out political system involving algorithmic data processing here. I do not even think any one answer can ever be fully convincing and appropriate for all systems.

I merely wished to show that embracing total digital surveillance, and the loss of privacy which goes along with it, can be in tune with affirmation of the norm of negative freedom, and that it is in our interest to implement such structures, as this seems to be the most promising way of using digital data in a democratic manner and not in a way that it primarily serves the interests of governments or private companies. In a way, this would be a European social credit system which included a democratic usage of our digital data.

Comprehensive data collection can also replace existing elements which fulfil the role of a social credit value with more reliable ones, such as Schufa, which is a German private credit bureau, criminal, or police record, official medical examination for candidate civil servants. This could result, for example, in socially inferior candidates becoming creditworthy through data capital. In contrast to the Chinese system, it should be based on the norm of freedom.

The more intensive the monitoring, the higher the probability of not committing a criminal or morally reprehensible act, but also the probability of sanctions for such actions (Sorgner 2017b). However, there is no perfect moral system. One danger associated with this is an increase in the number of problematic sanctions imposed for morally unacceptable acts. This is the real reason why we value privacy. In order to guarantee a plurality of lifestyles, we must promote freedom much further than is currently the case, and sanction only serious offences.

The state has a central role to play in this, and it should create the necessary regulations for the handling of data. If most information is collected by only a few private institutions, such as Google or Apple, it will be more difficult to guarantee the rule of law. Who has access to which data must be restricted. This insight is central, as human beings have a widely shared tendency to abuse their power, to use information in their own interest, and it is very difficult to undermine this risk. Consequently, it needs to be noted that in a system of total digital surveillance it ought to be primarily algorithms which have access to the digital data which is collected. Algorithms should primarily be responsible for monitoring the digital data. Only in exceptional circumstances, and in emergencies, should humans have the right to access the data. It is relevant to the preservation of the free rule of law that primary monitoring is carried out using algorithms and that human access must be strictly regulated, since the potential for abuse is undoubtedly enormous.

2.2.7 An as-good-as-it-gets ethic of reducing violence

So far, a number of reasons have been given for the pragmatic need for the collection of digital data and, together with it, the creation of a European social credit system. If the norm of negative freedom is further promoted, then we can build on key achievements in education and at the same time enable the expansion of health span, quality of life, and also financial well-being. However, when we speak of the norm of negative freedom (Sorgner 2010a) the concept of person must also be clarified, since freedom is a norm for persons. Traditionally, the status of a person has been granted exclusively to people. When other entities were included, these were usually angels and God (cf Sorgner 2010a, 2018b). However, this understanding of persons has been vehemently criticized most recently. Now we need to take a closer look at who should be granted the status of a person and what it means to recognize the norm of negative freedom for persons, which also has to do with the question of what harm is done to a person, because one person's freedom ends where another person's freedom begins.

Singer has convincingly shown why speciesism is morally problematic (Singer 2002, 2011a, 2011b). If two beings suffer to the same degree, then it would be morally wrong to view these two beings morally differently just because they belong to two different species. Hence the need to widen the moral status of personhood to beings that do not belong to the human species, to non-human animals, and possibly also to sufficiently developed AIs. Moreover, the separation of the moral status of personhood from belonging to the human species does not mean that a dualistic ontology is replaced on a legal level by a non-dualistic ontology. In order to avoid an ontological assessment by legal regulations, a secular legal basis must be created. Ontological judgements as part of the legal system are inevitably in contradiction to the foundations of a liberal-democratic system. Therefore, the necessity arises to exclude ontologies from a constitution as far as possible.

The proposal of a new basis of personal status is meant as an 'as-good-as-it-gets' ethics. We need a fictive proposal that covers many of our intuitions, but no more than that. The norm of negative freedom is a wonderful achievement. I affirm it, and I am glad that I share this insight with many other people today. I fight to strengthen its effectiveness, I promote the normative position, but I do not assume that it is ontologically better founded than any other moral position. Negative freedom implies the liberal ethics of a fictitious autonomy. If we think about it analytically, the theoretical challenge and pragmatic paternalistic implications of an ontological foundation of autonomy become clear. We are involved in many relational processes, events. and contexts. This does not mean, however, that relational ethics must be assumed. Such proposals are dangerous. Unfortunately, many critical posthumanists advocate a relational ethic without realizing that it restores dangerous totalitarian and paternalistic structures that must be avoided at all costs. Ethical nihilism is an important achievement that must be nurtured and developed. Ontologically, there are no standards. We are confronted with an ontology of continual becoming. For pragmatic purposes, however, fictive norms are necessary, and we can and must decide which ones we want to adopt. The central insight of the Enlightenment is that totalitarian and paternalistic structures must be avoided, since they necessarily suppress the plurality of human flourishing, and every human being has a unique way of flourishing.

Relational ethics transcend ethical nihilism and lead to the restoration of totalitarian and paternalistic structures. A concrete example from the Chinese social credit system illustrates this insight. The Chinese religious traditions, especially Taoism and Confucianism, contain relational ethics. Your role depends on relational structures. You are someone's wife, partner, son, or friend, and these relationships affect your rank in society. These structures have been integrated into the Chinese version of the social credit system. A real-life example illustrates the effects of relational ethics. In 2018, a Chinese high school student took a university entrance exam. He was successful and passed. However, he was not allowed to study at university because his father had failed to repay a loan.[24] The social credit value of the young man was therefore influenced by the father's behaviour and not by the actions of the young person himself. This is a necessary implication of relational ethical structures: your role and value depend on the actions of those close to you and dear to you. Your friends watch pornography or engage in unusual sexual relationships; this has an impact on your social value within the Chinese system. It is a necessary implication of a relational ethic that must be avoided. You want to be judged by what you have done and what you are responsible for. Are you ontologically responsible for your actions? Can autonomy be explained ontologically? These are not ontological questions. Ontologically, all these concepts cannot be meaningful. Autonomy is a fiction. But it is a fiction that has effects that are considered wonderful achievements. A person has the right to choose what she wants to do if it does not harm another person and if it does not violate the psychophysiology of another person. A person has the right to morphological freedom, that is, the right to choose what form she wants to take, develop, and establish. This is the best way to avoid violence and the emergence of paternalistic and totalitarian structures.

There are levels of personal status and degrees of autonomy. Autonomy is an ability. Education has the task of promoting the ability to make autonomous decisions. Becoming autonomous depends on the ability to abstract, to draw conclusions, to be able to talk about philosophical topics. Mathematics, languages, as well as creativity and logic are necessary to draw conclusions. Will we become independent of others by learning these skills? Ontological autonomy cannot even be conceptualized intellectually. However, life is about recognizing and holding on to one's own psychophysiological drives, affirming them, and living in accordance with them, which must be encouraged in the context of education. Cultural norms tell you that you have to behave in a certain way. Your psychophysiological drives have different requirements. You must recognize them and stand up for them. This is the tricky challenge that has to be mastered. It is an extremely difficult task that is a challenge for everyone. In this context, I have been able only briefly to point out the various relevant concerns.

2.2.8 Truthfulness, mindfulness, and impulse control as means to strengthen autonomy

So far, our cultural norms are being reproduced on social media platforms. Hence, the following concern has often been raised: does the use of digital media increase the likelihood of inauthentic decisions? There is a widespread concern that digital technologies via social media, advertisement, and politics are undermining one's own authentic interests. This thought is often explained further by reference to how Facebook was used during the 2016 US elections to influence the voters by means of personalized advertisements based upon a big-data analysis of each individual user. This social challenge is one which needs to be seriously considered.

The challenge for each one of us is to be ready to analyse oneself, understand one's own wishes, longings, and desires, and also to accept them. Even if one has some kind of understanding of what is in one's own interest, it can be difficult stick to one's own interest. This can have several reasons, for example, weakness of will, lack of impulse control, strength of external forces.

We all have our idiosyncratic desires, forces, longing, wills, and interests. At the same time, there are strong external forces which have created second natures in us. Sometimes we incorporate these into our being, sometimes we do not even realize the tensions between our second natures and more fundamental idiosyncratic drives, and sometimes we are afraid of living in accordance with what we long for, due to fear of external pressure, punishment, or some other kind of sanction.

Realizing which group of our idiosyncratic drives are most relevant to who we are is a daring and difficult challenge. Being able to distinguish between a central longing and a second nature which was created by external forces can be difficult. To be true to oneself, to realizing one's own interests, desires, and longings, and to be determined enough to stick to them is a complex and challenging enterprise.

Three virtues seem to be particularly relevant for strengthening one's own autonomy. All of them can be promoted by means of education, but they can be refined and developed further by means of technical enhancement procedures, too. In order to increase the likelihood of our making authentic decisions, it is in our interest to technically enhance these virtues, besides promoting them by traditional educative means.

The posthuman virtues I have in mind are truthfulness, mindfulness, and impulse control. Truthfulness stands for the basic stable attitude of being courageous enough to want to realize one's own drives, and being ready to accept them. Mindfulness analyses the variety of interrelated drives in which we are embedded so as to be able to realize the diversity of one's own interests. Furthermore, impulse control is supposed to guarantee that one is strong enough to persist in one's desires in the face of all the obstacles with which one is confronted while doing so.

Truthfulness implies the readiness, but also the capacity, to embrace the multitude of our forces. This can be a tricky challenge, in particular if one has grown up in a restrictive environment with rigid laws concerning good and evil. It might be the case that one already suspects that one's own longings do not correspond to socially accepted demands. Yet, whether one takes the stance of being truthful to oneself and accepting this stance is a further issue, and demands a lot of courage. Education can be a helpful tool for promoting truthfulness, for example by being confronted with literary texts in which internal struggles of a protagonist are dealt with, such that one realizes that one is not alone with one's own struggles. Scientific studies have also revealed the potential of certain psychotropic drugs for promoting truthfulness. If these drugs are taken in the appropriate setting and under psychoanalytical guidance, they can be a means for opening up to being truthful towards one's drives. The psychological and medical studies by Stanislav Grof (1975) are particularly noteworthy in this respect.

Mindfulness is a virtue which has its origin in Buddhism. As a consequence of the scientific work by Jon Cabat-Zinn (2017) it has increased in relevance in many aspects of our lifeworld; even Google has established a mindfulness centre, in which mindfulness trainers are trained, besides mindfulness meditations being offered to workers. Mindfulness allows us to analyse the relevance of various interrelated affects in which each one of us is entangled. Scientific studies have revealed that mindfulness can also be promoted further through technological means; for example, there are indications for nicotine further promoting the attitude of mindfulness. James Hughes gave a wonderful TEDx talk on radical mindfulness in Rome in 2013.[25] Therein, he analysed various pharmaceutical ways of enhancing mindfulness. It needs to be noted that there has been a long tradition of enhancing mindfulness technologically. The Japanese tea ceremony has been employed as mindfulness training for centuries. During a tea ceremony, three elements are particularly relevant: (1) the bodily practice of the ceremony; (2) the appropriate surroundings, the setting, the sounds; (3) the psycho-pharmaceutical effect of the tea which is consumed. Here, it becomes clear that it is implausible to regard psychotropic drugs as alien to mindfulness.

The third virtue which is relevant for promoting authentic decisions is that of impulse control, which seems to be particularly relevant for success, complex capacities, and the likelihood of living a flourishing life. Consequently, there have been a significant number of studies dedicated to the topic of promoting impulse control technologically. One drug, which was particularly successful in regard to impulse control, was that Tolcapone.[26] It must also be noted that various psychotherapies represent a promising method for dealing with the issue of impulse control.[27]

Obviously, these reflections merely represent a glimpse into the multifaceted and complex field of research on the technical enhancement of virtues. All I intended as part of these reflections was to show that the challenge mentioned at the beginning does not have to be crucial, that is, it does not have to be the case that the use of digital media will increase the likelihood of inauthentic decisions. Given that the virtues of truthfulness, mindfulness, and impulse control are promoted both by means of traditional education as well as by technical enhancement of these virtues, it can, rather, be stated that thus technologies can strengthen our autonomy, and together with this the likelihood of our living a good life.

2.2.9 Conclusion

I have explained that computers are getting smaller and entering our bodies so that we become upgraded humans who can interact efficiently with our environment in smart cities and have the means to cope with ageing, the world's worst mass murderer. This development is accompanied by new challenges related to digitalization. All these considerations make it a practical necessity to promote the comprehensive collection of data, especially big gene data, as we value both our personal prosperity and our economic prosperity and do not want to be worse off than our ancestors. If we are to promote our prosperity, we need to address these challenges and make the appropriate policy arrangements. In this way, our average health span can be significantly extended, which will further increase the likelihood of human flourishing.

Here I would like to emphasize that full monitoring does not have to mean giving up negative freedom. Total surveillance is not just a tool for the powerful to increase their power. At least, that does not have to be the case. There is, of course, the danger that the internet panopticon can be used in this way, which would confront us with a monitoring system of unprecedented intensity. This is the main reason why there is such a widespread fear of such structures. I can understand that; it is not that the fears are unfounded. One way of dealing with this fear could be to use primarily algorithms for monitoring. In this case, the likelihood of human abuse would be reduced. Psychological studies confirm that people prefer algorithmic to human judgement,[28] which seems to imply that humans are also less afraid of being monitored by algorithms than by other people. Algorithms can be programmed to take human rights into account. If humans are primarily responsible for monitoring, the risk of abuse increases. However, and this is another insight highlighted here, many of our personal interests can be promoted by big data analyses based on personalized long-term monitoring. Personalized data is necessary when it comes to gaining knowledge about the relationship between genes and health, genes and well-being, or life-style and human development. Large data correlations on all these topics require personalized data. The more data we have, the more correlations can be identified. Even if all these data were available, there would be many more moral challenges. Who should have access to which data? Who collects the data? Which goals can be promoted with the help of this data? Advertising flourishes particularly well due to data collection. Should advertisers also have the right to buy data from the government, if that is the institution responsible for data collection? At the moment, I assume that it must be the state that is responsible for collecting data, because any other organization or institution that has all this information at its disposal would soon become extremely powerful, which would give it enormous political influence.

This brief selection of questions shows how ground breaking the implementation of such structures is. However, we need data for economic well-being, for scientific research, for the promotion of well-being, and for the elimination of ageing. The achievement of all these objectives is so important that not collecting data is not a realistic option in practice.

With the help of a European social credit system based on recognition of the relevance of negative freedom we can further promote a wide variety of life-styles and the health- and welfare-related interests that depend on digital data collection. Whoever claims that these reflections are absurd might wish to consider the following statements.

- A pioneer in aviation, Wilbur Wright, said in 1901: 'I confess that in *1901* I said to my brother Orville that man would not fly for fifty years.'
- A father of radio, Lee De Forest, was still certain in 1926: 'So I repeat that while theoretically and technically television may be feasible, yet commercially and financially, I consider it an impossibility; a development about which we need not waste time dreaming.'
- A leading computer expert, Ken Olson, president of Digital Equipment Corporation, assumed in 1977: 'There is no reason for any individual to have a computer in his home.'

These remarks[29] could give us food for thought.

2.2.10 Final thoughts related to the COVID-19 pandemic

The main goal of these reflections has been to show that privacy, total surveillance, and how we can hold on to the wonderful achievement of negative freedom are the most pressing issues of our times which have to do with digitalization. Unlike Musk and some of his friends who suggest that we should be extremely afraid of an all-powerful AI which will wipe out humanity, I regard this thought as an attempt to entertain us with fascinating philosophical reflections so that we will lose track of the central pressing moral issues that we should be concerned with, namely, how we can uphold and cherish the wonderful norm of negative freedom in a world which is continually becoming more digitized, so that smart cities as well as upgraded humans will be realized.

To put it in a nutshell: 'When choosing between health and privacy we should go for freedom. Or why Harari is wrong.'

Harari claims in his opinion piece 'The World after Coronavirus':

> Asking people to choose between privacy and health is, in fact, the very root of the problem. Because this is a false choice. We can and should enjoy both privacy and health. We can choose to protect our health and stop the coronavirus epidemic not by instituting totalitarian surveillance regimes, but rather by empowering citizens. (Harari 2020)

But Harari is wrong. To promote health effectively, large amounts of data are needed. The more data we can use, the more reliable are the resulting correlations of health and our behaviour, genes, and external influences. Such data are also needed for innovation, scientific research, and policy making. All these procedures are of central importance for a country. Extensive amounts of data are needed for the economic, health, and social prosperity of a country. However, it must be ensured that the data is used democratically. Such a structure does not yet exist at present. In the US, data is primarily collected by large companies. This turns them into quasi-political actors, which has the potential to undermine the foundations of free democratic societies. In China, data is collected by the government on the basis of values and norms that cannot be reconciled with the achievements of the Enlightenment. The structures currently prevailing in Europe undermine our strongest interests. Here the focus is on data protection. The possibility of achieving an extensive collection of digital data at a political level is also being undermined in practice. However, this also means that we are losing the opportunity to use this data in the democratic interest and to promote our health.

Collecting data is particularly important when it comes to the issue of health. If we have to choose between health and privacy, and we should choose health, because the majority of citizens identify an increased health span with a better quality of life. We have no reason to regret giving up on privacy. We do not cherish privacy, but we cherish freedom. We can have health and freedom, but not health and privacy. Why do we think that we need privacy? There are two main theories that explain this, the property theory and the sanction theory, and they are not mutually exclusive.

According to the property theory, data is our intellectual property and therefore an extension of ourselves. If governments or anybody else take our data, they seem to expropriate us. But is this necessarily the case? If data are our property that we can exchange for other goods to our benefit, such as health insurance, then this is not expropriation. Having universal health insurance is an enormously important achievement but keeping the system alive demands huge financial inflows. Using our data to partially compensate for this service is in the interests of a society.

Furthermore, the sanction theory states that we fear that the data being collected, stored, and used by a government could be the basis of sanctions against us. We fear sanctions. However, sanctions are necessary. If a murderer of an innocent child is caught and sanctioned, this is just and widely accepted by the society. We merely do not wish to be sanctioned for acts which should not be sanctioned, whether morally, institutionally, or legally. This is the crucial issue. The fear of such sanctions is also the reason why we fear our personal data being stored in one place. How can these fears be dispelled? First, we need to reduce the possibility of access to the data by humans, because the risk of abuse is too high. Data access should primarily be granted to algorithms. Only in specialized circumstances should humans have the right to access the data. This is a significant challenge which needs to be dealt with continually. Second, we need to become much more open and pluralistic. Only those acts should be punished where direct harm is being done to another person. Currently, this is far from being the case in many parts of the world, even in the most developed democracies, such as Germany, where, for instance, incest among consenting adults is prohibited. It is merely a contract among two or more consenting adults to have sex together. The government should not have the right to have any say in this respect. If the consequences of an act were morally relevant, then a government could force a couple to undergo genetic tests before getting married, as is the case in Saudi Arabia.[30] In this case, one could also forbid two genetically deaf people to have offspring together. This undermines the right of reproductive freedom, which is one aspect of the norm of negative freedom.[31] What should matter when two or more consenting adults have sex is whether they have freely agreed to do so or not. If they have freely agreed to do so, the government should not intervene, as is the case in Spain, where incest among consenting adults is legal. Third, we should promote e-governance to make decision-making processes concerning our data more transparent. A lot more needs be said on this issue, from the need to establish a well-functioning, rapid infrastructure to the need to realize reliable algorithms to reduce the need of human intervention. However, this is beyond the scope of the present book.

It is not necessary for total surveillance to lead to freedom, but total surveillance can be in our interests under the circumstances I have mentioned. Total surveillance must not be expropriation, but it can be the price we pay for something which the majority of people regard as an important determinant of the quality of life, namely our health. We pay with our data for our universal health insurance. Total surveillance does not have to lead to illegitimate sanctions, either. To reduce the risk of this happening we need to store our personalized data safely, and it ought to be primarily accessible by algorithms. Finally, we need to promote plurality much further. Each person should have the right to live in accordance with his or her idiosyncratic needs, wishes, and longings. To clarify the limits of our individual freedom, we need to have a clear understanding of what should count as harm being done to a person. A person's freedom ends where the freedom of another person begins.

If we observe these lines of thought and implement them legally, we can achieve a democratic use of data. To this end, it is necessary to reconsider the importance of data and our assessment of the collection of digital data, because we have to make a choice between health and privacy. And in that choice we should choose freedom.

Critics of this approach rely on a voluntary route concerning surveillance. However, the realistic implementation of an opt-in solution with respect to an app for digital data collection is based on several problematic assumptions. You need to have an enlightened public where an enormous amount of people are willing to participate and ready and willing to opt in. In particular, in countries with a long-standing liberal tradition this is an enormous challenge. There are many reasons why people might decide not to opt in, or why this option is not a helpful one: not knowing that one is coronavirus positive, digital exclusion, lack of understanding, lack of trust, fear of sanctions, laziness, joy of being a free-rider, lack of care, absence of a strong individual motivation. The availability of such an app might even worsen the situation, as it might also convey a false feeling of security, which might increase the likelihood of people taking fewer precautionary measures.

The central issue which needs to be taken seriously is that we already have clear empirical evidence for a lack of interest in participating in opt-in solutions, even though a specific act is seen as morally right. The case of becoming registered as an organ donor is structurally analogous to that of opting into a coronavirus app. In several countries with an opt-in for organ donations, donating one's organs is seen as the morally right thing to do, for example in Germany. Still, most people do not get registered to be organ donors, even though there is an enormous shortage of organs.[32] However, in countries with an opt-out, where donation by default is the general principle, for example in Austria, hardly anyone decides to opt out, as they regard organ donation as an ethically trivial act.[33]

In theory, an opt-in app might sound good, but the enlightened, altruistic, active citizen who is digitally literate and willing to actively opt in to such a system is far removed from the life-world. Getting registered on a coronavirus app is structurally analogous to getting registered as an organ donor, which is the main reason why analogous public reactions can be expected.

In addition, it would not take many people to undermine the system of opt-in registrations and cause an enormous amount of new infections. You just need to have a couple of asymptomatic people as super-spreaders, who do not opt in, in order to bring about many more infections in a community; for example, a bartender, an active member in a religious community, a healthcare worker.

Furthermore, the regulation which I suggested has the additional advantage that the digital data is also used to at least partially pay for the universal healthcare system, which is an enormous achievement. It takes seriously the relevance of data collection for innovations, scientific research, and policy making. It is in the interests of the people as well as of a government to be able to collect and use digital data. To guarantee that an enormous plurality of different lifestyles can be embraced in a society while a highly efficient universal healthcare system is available, we need a democratic usage of our digital data. This approach seems a promising initial step for developing appropriate social, legal, and political structures for realizing a proper democratic usage of our digital data.

2.2.11 Glocalization and the war for digital data[34]

The related topic of 'Glocalization and the war for digital data' deserves to receive further attention from the great variety of posthuman perspectives.[35] The following reflections seem to be particularly relevant.

The world has been becoming more and more economically globalized for several centuries. Trade between countries usually promotes peace. If you are dependent upon goods from another country, the likelihood of entering a war with that country is reduced, as you are not interested in losing the goods which the other country provides. On the other hand, there has also been a rediscovery of the local and the ways that global phenomena are imbricated in local practices, a process described as 'glocalization'. Hybrids between established local structures and global institutions are created; a mundane example from the business world is Starbucks offering a baguette in Paris or a focaccia in Milan.

Recently, the politics of digital data (Sorgner 2019a) has represented a new example that underscores a further development of what states have always been fighting for, namely, power. It is widely stressed that 'data are the new oil'.[36] Even though there are many dissimilarities between data and oil (one is intellectual property, whereas the other is a natural good), they are similar at least to the extent that both are commodities connected with state power. Glocalization starts to play a crucial role in this context, too, as we see that states aim at using the 'new oil' to increase their state power, but their approaches are local.

The first question in the new constellation is: who is able to access digital data? The answers to this question are local. In China it is the government, which has the possibility of total surveillance and thereby the collection of all kinds of data, including personal ones, in massive quantities. In the US, data are primarily collected by big companies. Yet, companies have access only to specific data, and they cannot force people to provide them with digital data. In the EU, on the other hand, rigid data protection laws have been introduced that undermine the possibility of mass-collecting digital data. However, big data are needed for all kinds of enterprises, from realizing technological innovations to political decision-making processes to proper scientific research, and therefore are centrally relevant for financial, individual, and bodily well-being.

The differences in data-collection practices among countries may have enormous consequences. Generally, states that embrace the collection and use of digital data will be the ones that can be expected to flourish economically and scientifically. Due to the actual usage of total surveillance, China has better possibilities in this respect than the US. It is already the case that China has published more papers in peer-reviewed journals than has the US,[37] and due to the continually increasing availability of data it can be assumed that this tendency will become even stronger.[38] In Europe, on the other hand, the legal structures restrict the possibility of a substantial collection of data, which undermines many European interests, such that a constant decline in the quality of life, economic success, and scientific progress can be expected.[39]

Unfortunately, Europe has also fallen behind because it missed important aspects of digital developments in rather unforgivable ways under both the EU and national governments years. The European digital infrastructure remains insufficient, the costs for the usage of digital data are too high, and innovative tech companies (in both hardware and software sectors) are still not sufficiently sustained.[40] Technological innovation in Europe is currently floundering, due to all the aforementioned mistakes and challenges, and not just due to the EU's privacy laws. However, if big data analysis becomes as relevant as is expected by, for example, the Organisation for Economic Co-operation and Development, the U.S. Committee for Economic Development, and other global organizations, then it is reasonable to expect that the economic implications for Europe will be devastating ones.[41]

East Asian countries, on the other hand, have embraced the use of emerging technologies for many decades, which has enabled them to develop rapidly.[42] Their economic success has been so significant that the cost of labour has increased to the point where more and more of it is reshored and moved to selected African countries.[43] This trend might be more beneficial to Africans than development aid from Western countries, as it has the potential to create sustainable and flourishing local infrastructures. Traditional Chinese religious mentalities such as Confucianism support the possibility of collecting digital data by subsuming the individual under the common good. This represents one of the main challenges for Western countries, and the US has certainly realized the economic, political, and scientific potential connected with the use of digital data in China. President Trump's sanctions against Huawei were an expression of a digital data war,[44] but Europe seems hardly to have recognized this reality. Purely in terms of the efficiency of collecting digital data, the libertarian US system cannot compete with the centralized Chinese model. Europe's social-liberal democratic system must find its own local ways to do so, and it is high time to get on track when 'glocalization' means that every area has to deal with the new opportunities and related global pressures of data collection in its own specific ways.

From the standpoint of open societies, there are many problems with this development. We used to believe that (aside from corporate firewalls for the protection of intellectual property) the internet was a globally open entity. However, China has shown otherwise: either you adapt to the Chinese system or you are banned from running your company or accessing the huge consumer market in China.[45] China has its own rules, and Russia seems to be moving in a similar direction.[46] Thus, rather than an open global data space, the internet is no longer available globally in a uniform way. Great firewalls can be erected that strengthen local structures, as the example of China shows. How will the US and Europe react to these developments?

As mentioned, the economic system is already globalized. Chinese companies such as Huawei are globally successful, enabling the Chinese government to collect digital data in liberal political systems all over the world. In addition, the Chinese government has also created structures in China that enable the Chinese government to collect Chinese digital data and that prevent external companies from doing the same. Still, collection of digital data within a country is significant only if the country is important enough in size. A Lithuanian firewall might not be too advantageous for Lithuania. This is a further reason why having the EU as a political unity is so important for all the member states. Italy, Lithuania, and Portugal are not big enough to be successful alone in a globalized world. If they were to negotiate with China or the US, Europe as a whole might get a good deal. Lithuania alone would not.

If the EU wishes to have a bright future in which health, wealth, and the sciences can flourish, European states need to stay together, but they also need to rethink the importance of digital data. If the earlier reflections are plausible, I do not see how a European digital firewall can be avoided. Europe also needs to collect digital data, as having this information is a prerequisite for glocalization and a European flourishing of health, wealth, and the sciences. However, these revisions must not and do not have to undermine the achievement of the norm of negative freedom, which is so centrally connected with the European Enlightenment process. Collecting digital data and affirming the norm of negative freedom are two independent processes, if the digital data are used in the public interest, for example, in the interest of European residents.[47] Most people identify an increased health span with a better quality of life (Bostrom 2009, section 3). If digital data are used by the EU to improve the public healthcare provided in Europe, then this will be in most residents' interest. Public healthcare systems are an enormous achievement. They are extremely expensive, too, both with respect to the people working in the system and in regard to technological innovations in the field of medicine. Europe has to enter the war for digital data if it wishes to keep and promote the quality of life it has. The second step Europe needs to consider is that glocalization processes demand that local interests require more attention than does the movement towards an increased global integration.

All these developments show that digitalization does not necessarily imply expanding globalization, but there is a need to rethink political structures by taking the relevance of digital data as well as the norm of negative freedom seriously. In a certain sense, globalization exists in the World Wide Web, inasmuch as we can access Chinese websites. The Chinese can also enter non-Chinese websites, but only if they access sites that are not censored by the Chinese government or if they privately use illegal VPNs (virtual private networks). This is the insight that needs to be considered carefully. Globalization has become restricted in this respect. So, rather than globalization, we should focus on the dynamics of glocalization, as cultural developments and relationships are dynamic and multidirectional.

This does not mean that the era of globalization is over. Pharmaceutical drugs are produced in China, and we are dependent on them. This is globalization. Yet the Chinese have also created a digital Chinese firewall. They are going local, and it seems to be a successful move for getting hold of digital data which other institutions cannot access. This forces the West to act accordingly. Europe needs to rethink the meaning of digital data and the consequences of its policies. This does not seem to have occurred so far. Germany is even considering allowing Huawei to be involved in realizing 5G in Germany.[48] This could have financially devastating consequences for Germany, as well as for the EU in general. The Chinese digital firewall is turning digital data into a local phenomenon. This development has significant consequences for glocalization, and European countries need to rethink their digital data policies and act accordingly.

3

On a Carbon-based Transhumanism

Genome editing might be the most important scientific invention of the beginning of the 21st century (Knoepffler/Schipanski/Sorgner 2007). Selecting a fertilized egg after IVF and PGD is a biotechnological option already. We can have children with three biological parents. Biobags enable the development of unborn lambs, which show the likelihood of artificial human wombs being created. Japan allows human–animal hybrids to be born. The possibilities of modifying humans and other animals by means of gene technologies are enormous. Yet, many of the procedures which are technically feasible are still legally forbidden. The decisive breakthroughs concerning CRISPR occurred only in 2012. The human genome was fully deciphered in 2003.

Decisive developments in gene technologies have happened at the beginning of the 21st century. They have the potential to enable us to enhance evolution. We can modify ourselves and actively realize specific traits. We might even be able to pass on some of these traits to our offspring. Yet there is an enormous hesitation to embrace these technologies in continental Europe. Even though there are fewer hesitations in the Anglo-American and the Eastern–Asian world, there is still widely shared uncertainty concerning the moral legitimacy of the great variety of technologies and associated possibilities.

It can be expected that developments in the field of gene technologies will accelerate further if the use of big data digital analysis concerning correlations between genes and traits is developed further. In the US, private companies are trying to promote this goal; for example, 23andme with its over 5 million clients. In Estonia, the government pays for gene analysis if the result is shared with the government. In Kuwait for a certain time period it was mandatory for all residents and visitors to submit samples of their DNA.

Gene analysis is the prerequisite for gene technologies, and I addressed various issues concerning digitalization and gene analysis in Chapter 2. Here, I will focus on the question how to deal morally with gene technologies. I will present a non–utopian Nietzschean transhumanism, which embraces a radically pluralistic concept of the good, and I will progress as follows.

I will start with an analysis of the cultural relevance of debates on gene technologies by analysing the cultural reception of some of these ideas. I focus on 'Rules for the Human Zoo' suggested by Sloterdijk (1999; 2009), which caused a massive debate. He presented these reflections in 1997 and 1999, and published them in 1999. Habermas' seminal essay on liberal eugenics from 2001 is an implicit reaction to Sloterdijk's reflections. Therein, Habermas stresses a strong connection between some of Nietzsche's ideas and the reflections of various transhumanists. He regards both of these as dangerous. I regard this as an implausible position.

Savulescu and Persson have stressed the need to realize and promote moral bioenhancement first, so as to reduce the likelihood of abuse connected to other enhancement technologies which promote positional goods. In the second section of this chapter I explain why this suggestion is neither plausible nor realistic. It presents one of the aforementioned unrealizable and potentially dangerous utopias. It is more promising to promote morality further indirectly, namely by supporting biotechnologies for realizing cognitive capacities. This is the reason why I regard it as appropriate not to focus further on moral bioenhancement technologies, but on the most promising field of enhancement technologies, that is, gene technologies, which includes gene modification as well as gene selection.

Section three is dedicated to dealing with the question of how to evaluate gene modification technologies, for example genome editing by means of CRISPR-Cas 9. Against Habermas, I argue that the process of genetic modifications of someone's offspring is structurally analogous to traditional parental education. Structurally analogous processes ought to be evaluated morally analogously, too. Hence, just as education can be morally appropriate as well as morally wrong, the same applies to genetic enhancement technologies, too. As long as parents do not abuse their children, the right of educational freedom should prevail. The norm of negative freedom is an enormously important achievement.

The other promising gene technology is that of selecting a fertilized egg after IVF and PGD. In contrast to Savulescu, who claims that we have a moral obligation to select and implant the eggs which have the greatest chance of living a good life after IVF and PGD, I regard this demand to have some problematic, violent implications. In section four I critically deal with this argument of his. In contrast to his utilitarianism, I uphold the relevance of a fictive notion of autonomy. In the same way that humans should be free to choose a partner (or partners) for reproductive purposes, due to the right of reproductive freedom, the same should apply to the case of selecting fertilized eggs after IVF and PGD. Further clarifications concerning my ethical stance will be presented in Chapter 4, entitled 'A Fictive Ethics'.

3.1 From Nietzsche's overhuman to the posthuman of transhumanism[1]

Transhumanists teach at leading universities, dominate academic discussions in specialized disciplines, and are being considered in popular culture too. The 2014 movie *Transcendence* with Johnny Depp as well as Dan Brown's 2013 novel *Inferno*, in which transhumanist issues are at the heart of the plot, are clear indications for transhumanism having gained wider popularity.

The question concerning the relationship of transhumanism and Nietzsche's thinking is still an open one. Habermas identifies these two approaches, as he knows that Nietzsche's philosophy is still seen in the context of fascist thinking, at least in Germany. Hence, he employs his judgement to discredit transhumanism. Some transhumanists are well aware of the danger of being discredited by being identified with Nietzsche's approach. Hence, they too have seen the need to dissociate themselves from Nietzschean thinking. The director of the Future of Humanity Institute at the University of Oxford, Nick Bostrom, is a prime example of this position (Bostrom 2005, 1–4). However, both Habermas and Bostrom are wrong. In a heated academic debate concerning this relationship, which centres on some of my articles in which I develop a weak Nietzschean (or posthumanist version of) transhumanism, this becomes clear. A leading defender of a contemporary version of transhumanism, Max More, who participated in this exchange of articles, makes clear that he himself was strongly influenced by Nietzschean thinking, which is a clear proof of Nietzsche's influence on transhumanist positions (More 2010).

3.1.1 Reflections on this challenge

Nietzsche was aware that his thinking is dynamite. The same applies to transhumanist reflections. The well-known American intellectual Francis Fukuyama calls transhumanism the world's most dangerous idea (Fukuyama 2004, 42–43). I think this is correct for people who affirm a traditionally dualist account of the world, like a traditional Christian or Kantian account of the world. It needs to be acknowledged and considered that many legal systems contain encrusted structures of our Christian cultural past. This can be clearly understood, given the importance of categorically dualist accounts of human beings on which most constitutions rest. A prime example is the

German basic law, which stresses that even though animals are not things, they still fall under object law. Only human beings possess dignity and are seen as subjects. This position rests on and affirms the traditional Cartesian relationship between human beings and animals. Human beings have a material body and an immaterial soul. Animals, plants, and dirt consist of matter only (Sorgner 2013a, 135–159). Such a world-view is also still affirmed by contemporary thinkers like Habermas. Even though he claims to hold a soft naturalism (Habermas 2004, 877), his position entails that human essence lies in something which cannot be analysed empirically. Even though Habermas does not claim explicitly what it is that lies beyond empirical analysis, it is clear that the traditional concept of the immaterial soul, consciousness, or personality is implicitly contained in this philosophical position.

Both Nietzsche and transhumanist thinkers criticize such a dualist version of anthropology and attempt to move beyond it (Sorgner 2013a). Hence, their approaches have to be seen as dynamite or as dangerous by people and thinkers who uphold a different concept of the world. It is the conflict between dualist thinking and non-dualist thinking which is centrally relevant for the impact of both approaches. Due to their non-dualist approaches, their views represent a challenge for traditional philosophical views from the Western philosophical tradition, as dualist thinking has been dominant in these approaches at least from Plato onwards, whose views represent one of the central cornerstones of Western thinking. Traditional Eastern religious thinking has been much more open to non-dualist approaches. Consequently, a more complex exchange with these traditions could be of great interest for both parties. It might be interesting to note that some transhumanist approaches bear a close affinity to Eastern thinking. Most noteworthy in this context might be the transhumanism upheld by James Hughes, who used to be a Buddhist monk. His next book will most probably be entitled 'Cyborg Buddha', and therein he plans to present one version of how these movements can benefit from each other. A close affinity to Eastern thinking can also be noted with respect to Nietzsche's philosophy. The collection *Nietzsche and Asian Thought*, edited by Graham Parkes (1996), represents a particularly noteworthy publication from which a comprehensive overview concerning this topic can be gained.

As Nietzschean and transhumanist approaches see human beings as part of this one natural world, integrated in dynamic processes and merely gradually different from other animals; it is plausible that they have come about solely as a consequence of evolutionary processes, and it is highly likely that human beings will have to continuously adapt themselves to their environment in order to survive. Either you die out or you adapt and survive. Human beings have always developed, used, and employed technologies in order to fulfil this task and to make their lives better, more fulfilled, and more efficient. It is this basic insight which is closely related to Nietzsche's and transhumanist reflections. Furthermore, given that human beings have come into existence, it is likely that they will also eventually develop further, provided that they do not die out beforehand. In Nietzschean words, human beings represent the connection between animals and overhumans. By becoming a higher human being, human beings can increase the likelihood of developing into overhumans. In transhumanist words, by working on themselves, human beings can develop into transhumans, such that the likelihood increases that a posthuman can come into existence. There is a clear structural analogy between higher human beings and transhumans and between overhumans and posthumans (Sorgner 2009).

The posthuman in transhumanism can be a silicon-based entity on a hard drive, as well as a carbon-based entity who is a further-developed human being. It can be further developed merely with respect to one quality while still belonging to the human species, but it might also be the case that it is developed much further so that one can talk about the coming about of a new species. Even though there is plenty of evidence that this is the appropriate way of understanding the famous *Zarathustra* quote in which human beings are described as a rope between animal and overhuman, quite a few Nietzsche scholars tend to be critical of this reading of Nietzsche's philosophy (Niemeyer 2016). Their main reason for doubting this reading of the overhuman is his critical remarks concerning Darwin. Yet they need to be conceptualized in an appropriate manner.

According to Darwin, human beings are dominated by a struggle for life, which Nietzsche understood correctly (Nietzsche, 1967 [hereafter KSA] 6, 120). Nietzsche, on the other hand, disagreed fundamentally with Darwin in this respect, as he sees the will to power as the most fundamental basis of all human acts (KSA 6, 120). Hence, Darwin and Nietzsche disagree fundamentally concerning the most basic source of motivation of human acts. Nietzsche recognized this point and hence had to and felt the need to distinguish his own approach from Darwin's. He acted accordingly, as he did whenever he came across a thinker whose approach was composed on the basis of a similar spirit but was different from his own approach. Nietzsche explicitly stresses this issue.[2]

Here, Nietzsche's criticism of Darwin is his way of stressing that Darwin's and his own anthropology are very similar in spirit. Both see human beings as non-dualist, naturalistic, and this-worldly entities, which are solely gradually different from other animals. They came about on the basis of evolutionary processes and will develop further on that basis too, provided that they have not already died out. Like Darwin, but less detailed and informed than him, Nietzsche attempted to develop his own evolutionary theory in his writings not published by himself.[3]

Given these reflections, together with his detailed will to power ontology, it is hard to doubt that Nietzsche's concept of the overhuman is an evolutionary one. Still, it is an open question which qualities can be attributed to the overhuman, and it is by means of this question that the scope of concepts of perfection which contemporary transhumanist scholars uphold can be faced. On the one hand, there is the classical ideal of perfection, and on the other hand a radically pluralist concept of the good, which can both be found in Nietzsche. There are several statements which show that Nietzsche is a defender of the classical ideal,[4] and that the classic style is the highest one (KSA 4, 178).

These statements gain further support in a negative way by means of his remarks concerning the revered cripple and the genius in *Zarathustra*.[5] Geniuses are reversed cripples, because they have too much of one characteristic, whereas regular cripples are characterized by him as lacking one thing.[6]

Of course, it needs to be noted that Nietzsche does not claim that human beings strive for this ideal because this concept of beauty corresponds to their own nature. He stresses that this ideal is born out of chaos and permanent conflict.[7]

A similar ideal can also be found in and is also often associated with transhumanist thinking. Bostrom, in an early article of his, upholds a similar position: 'Transhumanism imports from secular humanism the ideal of the fully-developed and well-rounded personality. We can't all be renaissance geniuses, but we can strive to constantly refine ourselves and to broaden our intellectual horizons' (Bostrom 2001).

Still, it is questionable whether this classical or Renaissance ideal is actually a plausible one within an evolutionary framework. It must not be forgotten that Darwin stresses that the fittest one survives. However, the fittest one does not have to be the strongest, most intelligent, and most beautiful one, because otherwise the dinosaurs would not have died out. Only those organisms survive which are adapted best to their environment. Yet, it is an open question whether the Renaissance genius always has to be best adapted to his or her environment. It might be that sounds on Earth will be so loud that only deaf people will survive. In other words, it needs to be noted that it is unclear which qualities will make us fit in future circumstances. Consequently, given the evolutionary framework, it is unlikely that a single idea of the good is in our best interest concerning the question of human survival. In addition, it needs to be noted that a focus on the individual's likelihood of living a good life ought to be central. Do we lead a good life only if we live the life of the Renaissance ideal? In any case, Nietzsche's anthropology also provides a reply for this worry by identifying the importance of following the needs of one's own physiology.[8]

A well-grown human being is someone who follows her or his physiological demands. It is this concept which becomes particularly relevant for Nietzsche when describing master morality, because this is what physiological masters do – they follow their instincts and physiological demands. In Chapter 4, section 2, I defend an analogous concept in transhumanist circumstances (Sorgner 2016b, 141–57). To live a flourishing life, it is necessary to follow one's psychophysiological demands, and it is highly unlikely that any non-formal account of the good can describe a plausible concept of the good, because human drives, wishes, fantasies, instincts, and hopes differ radically from one another. To acknowledge the plurality of concepts of the good on a legal as well as on a moral level, is not only in the interest of most human beings but also in our evolutionary interest, as it is not advisable to put all the eggs into the same basket, as we have learned before.

Given the earlier reflections, a great variety of ideals of perfection or concepts of the good which are associated both with Nietzsche's philosophy and with transhumanist approaches need to be noted. In Nietzsche's case, I suggested the following solution to the question of the good, as both concepts are in tune with his thinking, and both were held during the same time period of his life, his latest and mature way of thinking. In Nietzsche's case, it seems to me most plausible to hold that he did affirm that human physiologies differ radically from each other, but the classical ideal of the good is the one which corresponds to his own ideal, the one which he aspires to realize. These two views of the good are not in conflict with each other within his approach but follow from each other. By listening to his own demands, he realizes the need of the classical ideal for his personality.

3.1.2 Implications for contemporary discourses

What is more relevant for us today is the consequences these two concepts have for our everyday politics, ethics, and morality. In this respect, my central worry with any detailed concept of the good needs to be addressed. Any strong concept of the good is paternalistic and violent, especially when its validity has legal implications, because it does not adequately consider the great plurality of possibly flourishing lives. Ethical nihilism which affirms that any non-formal account of the good is bound to be highly implausible is too great an achievement for it not to be considered adequately. What consequences does the acknowledgement of such an ethical nihilism have? Firstly, ethical nihilism demands continual criticism of encrusted totalitarian structures; secondly, ethical nihilism rejects the necessity of transcending a nihilist society so that a new culture becomes established; thirdly, ethical nihilism demands promotion of institutional changes so that plurality is acknowledged, recognized, and considered appropriately on legal, ethical, and social levels. Even though the remarks so far are rather formal, it is possible to specify them further concerning issues of emerging technologies, whereby many specifications concern the act of procreation. The examples of incest, hybridization, three biological parents, and selection after IVF and PGD reveal how concepts of the good embedded in legal constitutions still infringe upon the autonomous decisions of adult parents concerning acts where no one is harmed.

I will make some brief comments upon each of the preceding cases. Incest among adults is legal in Spain but illegal in Germany. Recently, there was the case of a brother and sister in Germany who grew up separately, met as adults, fell in love and had children together. The state threatened to put the father in prison if they decided to have further children. The couple did not obey and the father sent to prison. These were merely two adults who were in love with each other and wished to have a family. No one was harmed. Giving birth to another human being cannot imply a harm, as harm can be done only to an already existent entity. Entities that do not yet exist cannot be harmed. Even though all of this is the case, the concept of the good which is part of the German legal constitution does not regard such a partnership as appropriate. A radically pluralist concept of the good acknowledges that such a partnership can be an appropriate one.

Hybridization as an issue came up when scientists in the UK decided to undertake research with human–animal hybrids or parahumans. Between 2008 and 2011, more than 150 such embryos were created, but they had to be destroyed after a certain period of time.[9] The question remains why it is necessary to eventually destroy such embryos. They do not represent a threat to human dignity, because the concept human clearly does not apply to them. Japan has realized this insight and allows human–animal hybrids to develop to full term. In addition, by means of hybridization the likelihood of our survival and our individual flourishing might eventually even be increased. The following research undertaken by Dutch scientists supports this suggestion.[10] They alter zebra fishes genetically such that they can photosynthesize and thereby acquire part of their nutrition. The fish turn slightly green as a consequence of this procedure. Zebra fish and human beings are genetically not too different from one other. Given population growth, and the issue of overpopulation,[11] the question of our limited resources is an urgent one. Maybe such modifications or hybridizations will be one solution to this challenge. In addition, given the planned Mars missions, such procedures might be one way of solving the question of providing a sufficient amount of nutrition for future space travellers. The little green human beings known from science fiction literature might even represent the future of our species.

The next topic which reveals how strong concepts of the good limit our procreative freedom is that of threefold biological parenthood. In the US a special type of infertility treatment called Cytoplasmic Transfer was legal for a limited amount of time, and by means of it between 30 and 50 children were born who have three biological parents. Now, a different method has been developed and the issue of three biological parents needs to be addressed anew. It concerns parents with specific mitochondrial diseases who wish to have biologically related children. Several techniques have been developed. One option is to remove the nucleus from the cell with the defective mitochondria and then combine it with a healthy donor cell from which the original nucleus has been removed. In this case, there is an egg cell with DNA from two mothers, as the nucleus contains DNA as well as the mitochondria from the cytoplasm of the other cell. This cell can be fertilized, to produce a three-parent child. This technique offers not only mothers with mitochondrial defects the possibility of having a biologically related child, but also lesbian couples, or a partnership consisting of two women and one man. So far, the technology has been applied successfully in the case of animals. In the case of humans, the egg cells have successfully been created, but had to be destroyed for legal reasons. Consequently, there has been an intense debate in the UK as to whether such three-parent babies should be permitted. As no one is harmed by the use of this technology, it was approved in the UK in February 2015, but only in cases where the mother has a mitochondrial defect. Yet, given its reliability, it should also be legal for the other cases mentioned, which also raises additional questions concerning marriage. If three adults have a biologically related child together, should they not also be granted the right to marry each other, if this is in their interest? Is not marriage merely a contract among consenting adults to take care of each other? Is it not in the interest of a government, for such contracts to exist between more than two people, as it would reduce the likelihood of the government having to take care of a person who was not doing well financially? In a marriage, the partners have to take care of one another financially. In Colombia the marriage of three men has already been legally recognized.[12] What more can be demanded, for a group of people to count as a family, than the willingness to take care of each other?

The final procreative issue I wish to address here is that of selecting a fertilized egg after IVF and PGD. Technologically, this option already exists, yet there have been massive debates in various countries as to how far these selection procedures should be legitimate. Their legitimacy varies significantly from country to country. One of the central issues in this respect is that of the moral status of fertilized eggs, which varies significantly in the different states of the world. I personally think that it should be up to the parents to decide what moral status is attributed to their fertilized egg, as it is metaphysically unclear whether it has a special status or not. Furthermore, in section 4 of this chapter I will show in detail that selecting a fertilized egg after IVF and PGD is structurally analogous to selecting a partner for procreative purposes and, in the same way that the government ought to respect the procreative freedom of its citizens when they are choosing their procreative partners, it ought to do the same when the choice concerns fertilized eggs (Sorgner 2014b, 199–212).

Each of the four topics concerning procreative options shows in what way a legally valid, strong concept of the good can influence the autonomous choices of adults, while no harm is done to anyone by using the technologies in question. People who wish to use these technologies are treated paternalistically, in a violent manner, and are hindered from realizing their deepest wishes and longings. It is highly questionable whether such regulations are in tune with the basic guidelines of a liberal democratic society, and hence I think a lot needs to be done in this respect in most countries all over the world.

3.1.3 Conclusion

In dealing with the question of the relationship between Nietzsche and transhumanism, I have tried to hint at various central contemporary challenges related to this topic. This is not merely a disinterested academic investigation, solely of historical interest; by reflecting upon the various issues involved in this exchange we are confronted with many pressing contemporary issues in the field of bioethics. In addition, not only are specialized bioethical issues addressed, but also the wider picture of these issues needs to be considered. Do we live in a Christian society or do we live in a liberal democratic one? And in what kind of society do we wish to live? Should it be the case that a dualistic metaphysics is connected to our constitutions, or would it be better if a liberal democratic constitution attempted to get rid of strong ontological implications of any kind, so that the great plurality of lifestyles, preferences, and world-views is acknowledged? In presenting these challenges, I merely wished to raise the various issues in question.

3.2 Moral (bio)enhancement[13]

I regard genetic enhancement, morphological enhancement, cyborg enhancement, and pharmacological enhancement as extremely promising fields of enquiry, in so far as they may promote intelligence, memory, the health span, or beauty – characteristics that often contribute to a good life (Sorgner 2010a). These enhancement technologies can also be seen as promising ways to enhance morality indirectly. However, the issue is different with direct moral bioenhancement (Douglas 2011), particularly when it is supposed to solve the increased potential destructiveness of contemporary bio- and other technologies.

In their collaborative work, Ingmar Persson and Julian Savulescu suggest the need for moral bioenhancement, claiming that increasing technological possibilities will promote the likelihood of humanity's destruction, since there might be madmen who will use advanced technology for this purpose (Persson/Savulescu 2012, 46–59). Their main foci are not methods of indirect moral enhancement, such as education and promoting cognitive capacities. They affirm the use of these technologies too, but do not seem to regard them as effective. In a paper from 2013 they claim, 'the degree of moral improvement since the time of Confucius, Buddha and Socrates has been … small in comparison to the degree of technological progress, despite moral education', which seems to imply the ineffectiveness of traditional ways of promoting morality (Persson/Savulescu 2013). Consequently, they are interested in finding more effective means of promoting morality and they focus on direct moral bioenhancement. In an earlier paper they even argue in favour of moral bioenhancement that would be 'obligatory' – by analogy to 'education or fluoride in the water'. As they express it, 'safe, effective moral enhancement would be compulsory' (Persson/Savulescu 2012, 174).

I am more hesitant about moral bioenhancement, because I regard it, for practical purposes, as a highly problematic technology. In any case, I do not regard it as an option for dealing successfully with the increased potential destructiveness of contemporary technologies within a short-term framework, that is, within this century. In what follows I will explain why this is the case, and why, contrary to Persson and Savulescu, I think this need not concern us too much. In sub-section 1, I will critically analyse moral bioenhancement by means of citalopram, aimed at reducing the tendency to harm others directly. I give prominence to this method because it could be a practical option for altering human emotions and dispositions (Crockett et al 2010b) – to my mind, it seems the most promising practical option in this field.[14] In later sections I will consider other means of realizing moral bioenhancement, since the first option does not do the job it is supposed to and because Persson and Savulescu are concerned with alternative approaches.

I propose to show that any direct moral bioenhancement procedures that could be realized within a relatively short period of time are not realistic options. This does not have to concern us, however, because alternative options for promoting morality are available.

3.2.1 Moral bioenhancement by means of citalopram

In what follows I treat moral bioenhancement as related to the reduction of direct harm done to individuals, as this seems to me the most promising approach to altering human emotional tendencies, especially in light of Molly Crockett's research on serotonin (Crockett et al 2010a, 2010b) and the heated debate it inspired. Crockett gave citalopram to research subjects, altering their level of discharge of serotonin. Then she and her colleagues checked how the level of serotonin affected their reaction to the 'fat man' version of the notorious Trolley Problem (Thomson 1976, 204–217) and the Ultimatum Game (Güth et al 1982, 367–88). According to Crockett and her team, the results show that the higher the level of discharge of serotonin, the lower the inclination of the test takers to inflict harm directly on other individuals. Even though the effect described is a mild one, it has promoted considerable discussion around the possibility and desirability of moral bioenhancement. If it is seen as morally wrong to inflict direct harm on individuals, then a lower inclination to do so can be interpreted as a moral enhancement. If technological intervention to that end works in principle, then the likelihood is high that further research will lead to more efficient methods.

Doubts have been raised concerning Crockett's interpretation, for instance those of Harris and Chan (Harris/Chan 2010). They have criticized Crockett for identifying as a moral enhancement a reduced inclination to sacrifice a fat man to save five other human beings who would otherwise be killed by a runaway trolley. As Harris and Chan make clear, from a utilitarian standpoint this does not count as a moral improvement. They also point out that if Jasper Schuringa had possessed a higher level of discharge of serotonin, he probably would not have stopped Umar Abdul Mutallab, the so-called Underwear Bomber, from setting off a bomb on 26 December 2009. Schuringa would have had a reduced inclination to harm an individual, thus inhibiting him from acting against the bomber. But stopping the bomber was clearly the morally justifiable course of action. In the following sub-sections, I will deal with two options for implementing this type of moral bioenhancement: legal obligation and personal free choice. Each of these options has problematic implications.

3.2.2 Moral (bio)enhancement as legal obligation

One way to argue in favour of moral bioenhancement as a legal obligation is by analogy with vaccinations. Even though this analogy works in many respects, it fails with respect to a central premise. Health is a state that is in the interest of most human beings. This does not apply to a reluctance to directly harm an individual human being, the sort of bioenhancement that we are considering at this point. Indeed, under some circumstances harming an individual could be a morally justifiable act, maybe even a politically necessary one. To guarantee the state's inner and outer security, a police force and an army are needed; within certain limits, these have the right and the need to use force against individuals. The same applies to other people who find themselves in situations where force is needed; the case of Jasper Schuringa and the Underwear Bomber is just one extreme example.

There are also many everyday examples, such as the need to use force against violent youth who act aggressively against the elderly or the vulnerable. Such examples show why it cannot be in the interest of a country to render moral bioenhancement technologies legally compulsory, because this might, in the end, endanger the very existence of the state. For example, non-morally enhanced citizens from other countries might visit State Y, where it is compulsory for citizens to become morally bioenhanced. There could, accordingly, be outsiders who endanger the internal order of State Y by violent acts, fraud, robbery, or other crimes. State Y could also risk being invaded by rival countries whose soldiers are not morally bioenhanced. Again, morally bioenhanced soldiers with a heightened tendency not to harm other individuals directly would not be able to fulfil their duties adequately, thus possibly endangering the state's existence. Due to the risks to inner and outer security associated with moral bioenhancement technologies, living within a legal jurisdiction where these technologies are legally compulsory is not in the interest of most human beings. Leading a good life usually requires a safe environment; a state in which inner and outer security are at risk does not provide such an environment.

3.2.3 Moral (bio)enhancement as free choice

Would voluntary moral bioenhancement be of interest to individual consumers? The easiest case would be one where somebody simply wished to reduce her tendency to harm others directly. However, I doubt that this would apply straightforwardly to all, or even many, people, since acts of harming are so widespread and can be important means to protect a variety of human interests. One interest individuals could have in being morally bioenhanced would be to avoid sanctions for immoral and illegal behaviour. This case is well represented by a repeat criminal offender, for example, a serial rapist who is offered freedom if he is willing to undergo a form of moral bioenhancement such as chemical castration. The motivation of the offender to choose moral bioenhancement is to avoid the sanction of imprisonment. Even so, the procedure is not necessarily in his interest,[15] since sexual intimacy is central to his well-being, and he is thus sacrificing a part of his well-being to avoid being imprisoned. It is even possible to imagine violent offenders who connect acts of brutality with their personal conceptions of a good life. This is in no way a defence of their anti-social behaviour, but it vividly illustrates that there can be, and usually is, a gap between the moral life, judged from outside, and someone's idiosyncratic conception of a good life. While acknowledging the validity of this insight, we can, of course, also affirm that living in a society means that one has to limit one's possibilities for living as one chooses.

Alternatively, what about the case of a religious believer, Z, who is threatened with eternal torture if she acts immorally? Z holds that, by conforming to her faith's moral requirements, she can increase the likelihood of having a blessed afterlife. In her current life, however, when she feels she is being unfairly treated she lashes out with attacks that are (let us stipulate) legally and morally forbidden. Accordingly, she decides to reduce the risk of being sanctioned by the criminal justice system and especially in the afterlife by undergoing moral bioenhancement. On the one hand, her aggressiveness is positive for her because it proves she is strong and, therefore, it is connected to her feeling of well-being[16] (see Sorgner 2010a). At the same time, her aggressive acts are forbidden. Her ideal of a good life – one involving mastery and strength – is essentially connected with her aggressive inclinations. This ideal is curbed by living in a society and by her religious understanding that she will be tortured in the afterlife for aggressive behaviour.

It seems clear that some people might decide to undergo moral enhancement to avoid punishments. In Z's case, her behaviour can lead to being punished in both this and (at least according to her religion's teachings) the next life. By conforming to moral requirements, she can reduce the likelihood (as she understands it) of both secular and eternal punishment. If there were no political, moral, or religious sanctions, people like Z would not be interested in moral bioenhancement, because their morally questionable acts are also ones that provide them with intense feelings of fulfilment.

But could the type of moral bioenhancement under discussion be of interest to individuals who are not afraid of being sanctioned, people who operate within the law, but who are not interested in morality because they regard anything legal as also morally legitimate? One example could be a sadomasochist couple. Their neighbours might fear them, whether rationally or not, but the couple find sadomasochism fulfilling. Moral enhancement aimed at inhibiting impulses to inflict direct harm would (most probably) not be an option for them: given their proclivities, it would clearly decrease their capacity for living a good life.

Bioenhancement might become an option, however, if the drugs in question were effective only against causing harm in a more discerning sense – that is, if the operation of the enhancement drug distinguished situations where inflicted pain was not necessarily seen as *a harm*. Citalopram is probably not a drug that could be used for such a purpose, as it does not seem to be capable of such a sophisticated distinction. Will we be able to develop a drug that stops people from harming others – understood in a suitably discerning way – but does not stop them, in all cases, from inflicting pain? Remember, too, that in some cases it might be morally justified, or even obligatory, to harm others, as in the case of stopping a terrorist bomber (or, arguably, in war). Would not a drug be needed that is capable of distinguishing between morally justified and morally forbidden acts of harming other people?

Is it necessary to refer to such extreme cases in order to claim that this type of moral bioenhancement is not in the interest of most people, and hence would not be chosen freely by them? I doubt it. I think that it is important to realize that for *most of us* there may be a gap between our conception of a good life and any moral demand that we reduce our tendency of directly doing harm to other individuals. The following three brief examples exemplify how important directly doing harm to others can be in our everyday lives. First: suppose that, as you are trying to board a bus, some rude young men attempt to push past you even though they were behind you in the line. Here a certain verbal force directed at them is appropriate. If they succeed and you miss your bus, it will be very upsetting – so you resist their action. This is a very simple example to support my basic view that morality (in the sense of inhibition of aggressive or directly harmful conduct) often is in the interest of the immoral: for example, the rude young men succeed in getting into the bus by pushing past you, while you are left having to wait for the next one.

Another example: if there were an increased tendency to avoid directly harming any living thing, some valuable practices, and their associated professions, might vanish. If moral bioenhancement promoted a reluctance to kill fish for food, for instance, this might eliminate the profession of fishing and be contrary to what many fish eaters regard as a contribution to their leading a good life. Or imagine a social event where you see an attractive person whom you wish to meet – but your more assertive and talkative friend manages to push you aside and monopolize the conversation. It is his aggression toward you that leaves you alone and disappointed.

All three of these everyday examples reveal the tension between leading a less aggressive life and the importance of direct aggression toward other living beings, something that often helps you to maintain your own idea of a good life. I am not assuming a strong and detailed conception of the good by making this judgement. I am merely showing that a certain standing – a position of power and leadership – is a relevant component in most conceptions of the good life. Moral bioenhancement of the sort under discussion – a reduction of the inclination to do direct harm – is therefore problematic.

After reflection on everyday examples of human social intercourse, it becomes clear that there will be a clash between this understanding of morality and many, if not most, conceptions of a good life.

Of course, undergoing this type of moral bioenhancement might result in someone's conception of a good life itself becoming altered. This granted, it might also turn out that one's conception of a good life remains the same – even while the prospect of living such a life recedes. If that experience were commonplace, as I suspect it would be, the bioenhancement procedure would decrease the likelihood that many people would reach the goals essential for their conceptions of a good life. As I am assuming that a good life is immediately connected to someone's psychophysiology and that this differs radically among different human beings, any 'thick', non-formal account of the good is bound to be implausible. One aspect relevant for many people, however, is their will to live, to develop their capacities, and to attain some standing in their communities. Since these elements are components of the good life as conceived of by many persons, it is highly likely that there is a tension between a morality of not directly harming and many of the conceptions of a good life that are specific to individuals.

3.2.4 Further moral bioenhancement options

Persson and Savulescu do not employ not harming other persons as their sole criterion of morality. They also analyse the relevance of enhancing pro-social behaviour more generally, enhancing a sense of justice and reducing cognitive biases toward near-future outcomes (Persson/Savulescu 2012, 105–110). If we adopt their approach, a better practical example than citalopram might be oxytocin, which can be used to promote pro-social behaviour. As with citalopram, however, we can easily identify examples where the drug does not promote morality: examples that illustrate how pro-social behaviour cannot be identified with morally good or justifiable behaviour. Perhaps, for example, followers of Pol Pot represented *pro-social* behaviour of a kind, but it would be implausible to refer to it as, in any sense, behaviour that was *morally good or justified*. The same argument could be made by reference to members of many criminal organizations or totalitarian regimes who act pro-socially toward members of their own group – since these are the ones near and dear to them – but whose actions are nonetheless morally deplorable or even monstrous.

In a different respect, oxytocin is nonetheless superior to citalopram: even though pro-social behaviour can be an obstacle to leading a good life, this is less likely than in the case of not harming another person. Pro-social behaviour, belonging to a community, and having close intensive human ties are elements that are identified with a good life by a great percentage of human beings. Psychological studies support the plausibility of connecting pro-social behaviour and happiness (for example Aknin et al 2015).

Still, it needs to be noted that oxytocin has only a temporary, reversible effect upon human beings. Someone could take it in order to deal with a special situation, for example, a woman who is not able to love her child after having given birth. Taking oxytocin is not, however, a realistic option for dealing with the grand historic goal that Persson and Savulescu have in mind, that of reducing the risk of the global destruction of humanity. Criminals are not usually willing to take oxytocin when they feel the wish to kill someone. Hence, a more reliable technology would be needed. Still, it is an open question whether technologies with long-term, reversible effects (for example by means of specific drugs), or technologies with permanent, irreversible effects (for example by means of genetic modification), would be the better option. To get a better understanding of the implications, maybe we need to imagine possible outcomes that could count as successful direct moral bioenhancement.

If it were possible to genetically engineer some moral virtues, a different, more open attitude toward direct moral bioenhancement might be plausible. Let us say that scientists find out that there is a necessary correlation between a specific gene and the virtue of justice. Everyone with gene X would embody the virtue of justice, would regard freedom, equality, and solidarity as norms, would act upon them, would watch out when they are being attacked (and respond in an appropriate manner), and would try to convince others to act accordingly. Given such a situation, it might be possible to say that a Stoic Sage 3.0 had come into existence. If she were created in this manner, through a form of genetic editing, then I would – I think – be among the first to consider options for promoting the presence of gene X. In this case, several practical options might be available: wider use of genetic modification technologies such as CRISPR-Cas9 or using gene selection after IVF and PGD. Given that the methods for promoting gene X were sufficiently reliable, it might even be worth discussing whether it could and should be legally compulsory for everyone living within the boundaries of State Y to have gene X. In those circumstances, it would be important to consider the suggestions made by Persson and Savulescu (2008) when they argue in favour of obligatory moral bioenhancement ('safe, effective moral enhancement would be compulsory'). In such a situation I can imagine good arguments in favour of a structural analogy between compulsory education and obligatory moral bioenhancements by means of genetic interventions.

The problem with this example is that contemporary scientific research is far away from detecting any gene X with which the virtue of justice is identified. It is not clear whether, in principle, such a gene can even be conceptualized. Even if such a gene could be identified and detected in particular cases, realizing it in practice would be a further biotechnological challenge. As I have mentioned, gene selection after IVF and PGD could be an option, but it might not be possible to find gene X in all sets of fertilized eggs. We are, of course, even further away from realizing gene X in a reliable manner by means of genetic modification.

If we had a biotechnological way of promoting the virtue of justice, then I would value this moral bioenhancement procedure in the same way as I do moral education. I must also stress that I am not excluding the possibility that direct moral bioenhancement can eventually be made to work. Still, I do not currently see any way of realizing it within a short time frame and so I am extremely hesitant about the practical relevance of much of the recent academic discussion of moral bioenhancement (to which, admittedly, I am contributing with these reflections). Recall that moral enhancement is supposed to relate specially to reducing the risk of a technological destruction of humanity. To reduce that risk, however, enhancement of the virtue of justice would have to be enforced globally. This, in turn, would require a very far-reaching and powerful enforcement mechanism; the very idea of global enforcement seems to presuppose a world state, a global government, and moral norms that are globally shared. It is highly questionable whether the presence of such a global authority would, indeed, promote human flourishing, but I will not address that topic here.

So far, it seems clear that the most practical technologies – such as the use of citalopram – cannot fulfil the grand historic task assigned to direct biotechnological enhancement by thinkers such as Persson and Savulescu. Versions of direct moral bioenhancement that could plausibly do so can scarcely even be conceptualized, and they are technologically so challenging that they will not represent a realistic option within any short-term framework.

3.2.5 The relationship between cognitive and moral development

Given the general validity of the preceding analysis, along with the insight that direct moral bioenhancement is not one of tomorrow's hottest and most realistic technologies, must we infer that we are doomed – and that, given the developmental speed of powerful technologies, criminals or fanatics will soon be able to use them to destroy humanity? I do not think so, as I am in agreement with Harris's (2011) position that a betterment of our cognitive capacities also promotes morally good conduct. In order to increase the likelihood of a morally better way of dealing with the various challenges related to the application of emerging technologies, it seems plausible to expand cognitive capacities directly, instead of using moral bioenhancement, and I will now offer some arguments to this effect.

Earlier, I mentioned that Persson and Savulescu (2013) hold that there has been only small moral progress since Confucius, Buddha, and Socrates – at least when compared to the rate of human technological progress. I disagree with their judgement. On the contrary, it is feasible that, at least on a social level, cognitive and moral advancement have taken place in parallel with each other and with advances in technology.

Morality, as I understand it, is related to the recognition of norms such as negative freedom and human equality (Sorgner 2010a, 239–44) – norms that developed during the Enlightenment. Accordingly, when I consider the relationship between cognitive capacities and morally good or right conduct, it is these norms that I have in mind. How is it possible to think that these norms were in practice and respected in past ages without having been articulated consciously? If we held that there has been hardly any moral progress in this sense, imitating the approach of Persson and Savulescu,[17] we would have to assume that something like this was the case.[18]

Though I relate the notion 'morality' to norms such as freedom and equality, this understanding of the concept deserves further attention. Clearly enough, morality is a multifaceted notion that can be related to the concept of the right as well as to the values and norms in use during a certain cultural epoch. Furthermore, the notions of a good life, moral rightness, and their relation to the concept of morality might require further clarification. By directly identifying morality with norms such as freedom and equality, I implicitly signal that this is the notion of morality to which I subscribe and that I am using a particular notion of morality that is widely accepted within the German philosophical tradition. My use of the concept does not imply that there was no conception of morality or morally justifiable conduct before the Enlightenment. It does, however, imply that I am employing a contemporary understanding of morality as a criterion for judging whether and (if so) how far an arc of moral development has taken place.

Historical studies present us with reasons why the Persson and Savulescu thesis is highly implausible, since history makes it clear that neither the norm of negative freedom nor that of equality between all persons played any role in ancient Greek and Roman societies – or in the various kingdoms of medieval Christendom (Eissa/Sorgner 2011). In classical antiquity, political systems were strongly hierarchical. People could take it for granted that there was such a thing as a 'natural' slave; even the greatest ethical thinkers of ancient times, such as Aristotle,[19] agreed with this assumption. Much the same applies to the Holy Roman Empire during the Middle Ages. How can one even imagine that the moral and political norm of negative freedom played any role in communities before the Enlightenment? Citizens were forced to follow the religious, metaphysical, and moral beliefs of their religious and political rulers; Socrates was, after all, sentenced to death because of supposed disrespect to the gods.

Prior to the Enlightenment, political and religious authoritarianism and repression were common. Since the advent of the Enlightenment in Europe in the late 17th century, thinkers, scientists, writers, soldiers, and ordinary citizens have sustained a struggle for the right to live according to their own conceptions of a good life. It is primarily *this wish* – the commonplace, yet radical, wish to live a good life according to one's own criteria – that explains the central Enlightenment developments. The French Revolution, the increasing relevance of the natural sciences, and the development of new technologies by engineers relate closely to the fight against the dominance of Christian religious and aristocratic political leaders – the dramatic and ongoing struggle that made it possible to move away from totalitarian systems and towards modern liberal democratic political structures. On an intellectual level, philosophers from Descartes through Locke and Kant supported this development, as did artists and writers such as Leonardo da Vinci and the Marquis de Sade (Sorgner 2010a, 239–242).

Besides the genealogical reflections mentioned here, there is empirical evidence in favour of genuine social progress since the European Enlightenment. The psychologist Steven Pinker, especially in his monograph *The Better Angels of Our Nature*, has brought together empirical evidence showing that we 'orient [ourselves] away from violence and toward cooperation and altruism' – and that such a process has unfolded over millennia (Pinker 2011, xxv).

A reduction of violence against individuals implies a strengthening of respect for individual integrity. Various examples that Pinker discusses strongly indicate that significant moral progress has occurred. Pinker also stresses the relevance of empathy and reason among our 'Better Angels' (Pinker 2011, xxv) and the importance, in this context, of 'the concept of human rights – civil rights, women's rights, children's rights, gay rights, and animal rights' (Pinker 2011, xxv). Yet, when trying to explain the phenomena in question, he refers to a great variety of reasons, trends, and processes. Even though his reflections do not provide us with a final explanation of the processes that he identifies, his empirical data support the thesis that moral progress has taken place. If one grants that there has been significant moral progress from the Middle Ages to our time, affecting many citizens of enlightened and some non-enlightened countries, it might be related to the advancement of our cognitive capacities.

I regard it as highly credible that such advancement has occurred. There has been incredible progress in technologies, the sciences, and mathematics during the previous millennia, associated with a need to develop further and to gain more specific and more detailed cognitive capacities for dealing with this advancement physiologically. Furthermore, there is also some empirical evidence for cognitive advancement. One recent aspect of this development is referred to as the Flynn effect, which implies that there has been an increase of intelligence in industrial countries from 1930 onwards, given the results of standardized tests that have been in use since then (Flynn 2012).

If a cognitive advancement (as suggested by the Flynn effect) and moral progress (such as Pinker describes) have both occurred, moral and cognitive advancements seem to be correlated processes. Pinker refers to the Flynn effect as a potential reason for moral progress (or at least a decline in violence) and mentions the likelihood of a 'moral Flynn effect':

> We can now put together the two big ideas of this section: the pacifying effects of reason, and the Flynn Effect. We have several grounds for supposing that enhanced powers of reason – specifically, the ability to set aside immediate experience, detach oneself from a parochial vantage point, and frame one's ideas in abstract, universal terms – would lead to better moral commitments, including an avoidance of violence. […] Could there be a moral Flynn Effect, in which an accelerating escalator of reason carried us away from impulses that lead to violence?
>
> The idea is not crazy. The cognitive skill that is most enhanced in the Flynn Effect, abstraction from the concrete particulars of immediate experience, is precisely the skill that must be exercised to take the perspectives of others and expand the circle of moral consideration. (Pinker 2011, 656)

Pinker tries to explain the expansion in moral consideration of others by means of the increased prevalence of the cognitive skill of abstraction. It may, of course, be true that this line of thought is a mere suggestion, rather than an empirically grounded scientific explanation. For their part, Persson and Savulescu (2012, 107) acknowledge and emphasize the relevance of the work done by Pinker and Flynn, and there is an enormous amount of empirical evidence for at least a *correlation* between cognitive and moral advances. Although correlation does not equal causation, analysing historical developments suggests two relevant conclusions. First, contemporary human beings seem to be cognitively advanced, displaying an enormous capacity for grasping complex relationships, understanding many lines of causation, and developing unprecedented forms of technology. Second, given our current understanding of morality, there clearly has been moral progress. Nowadays, freedom and equality are far more widely shared in the world than ever before. Is there a causal relationship between the cognitive advancement and the moral progress that we can discern in human history? Although I cannot prove this, there seems to be at least a widespread correlation between them, and there may be a causal connection.[20]

Further scientific investigation is needed before we can claim with confidence that cognitive development actually promotes moral development. But one reason why it may be the case is that our capacity for imagining what it is like to be someone else promotes the likelihood of recognizing 'the other' as someone worthy of moral consideration. In all, there is some plausibility to the position that further cognitive enhancement processes will lead to further moral enhancement on a social level; and, contra Persson and Savulescu, cognitive enhancement does not have to be feared.

Thus, emerging technologies that further promote our cognitive capacities may also tend to promote morality. It may be true that the risk of global human extinction increases with emerging technologies, but the propagation of morality may also be promoted through technology. In that case, we do not have to be doomed, even if we are not in a position to enforce moral bioenhancement through global laws. In short, emerging enhancement technologies can be seen as indirect tools for promoting moral progress.[21]

3.2.6 Conclusion

We should conclude that direct moral bioenhancement is not a technological option for the time being. Perhaps, however, this is not so much a cause for concern as Persson and Savulescu suggest, since indirect moral bioenhancement, via cognitive enhancements, may work more effectively than they are willing to acknowledge.

In this chapter I have dealt with the question whether moral bioenhancement could be an appropriate means to the central goal that Persson and Savulescu regularly stress: that is, whether it is a plausible means to reduce the risk of human extinction. As part of their analysis, Persson and Savulescu point out that it is easier to do harm than to do good, and that (given the developmental speed of enhancement technologies) it will be even easier in the future for madmen to bring about human extinction through the use of powerful emerging technologies. Persson and Savulescu conclude that it is important to promote the morality of human beings not solely through education but also – and in particular – by means of technologies that directly promote morality. My response is as follows. We have reason to claim that promoting cognitive capacities and rationality by means of emerging technologies is sufficient to increase the likelihood of human beings acting morally on the social level. By contrast, it is highly unlikely that moral bioenhancement will do the trick within a sufficiently short time.

Given the correlation between cognitive capacities and morality on a social level, there are reasons to trust that cognitive enhancement techniques will reduce the risk of human extinction. If we employ genetic enhancement, morphological enhancement, cyborg enhancement, and pharmacological enhancement – all aimed at promoting general intelligence, memory, or the capacity to concentrate – we can produce significant moral progress, shaping a future in which the human species does not have to be doomed. Contrary to the worst fears of Persson and Savulescu, our species can continue to flourish and undergo moral development. Emerging technologies aimed at cognitive enhancement are promising, though admittedly indirect, means to enhance morality and the prospects of human survival.

3.3 Gene modification[22]

Habermas has put forward a strong argument against the position that educational and genetic enhancements are parallel events (Habermas 2001, 91). In response, I will provide reasons in favour of the position that there is a structural analogy between educational and genetic enhancement by modification, such that the moral evaluation of these two procedures ought to be viewed as analogous (Habermas 2001, 87). I will show that an affirmation of educational enhancement suggests an affirmation of genetic enhancement.[23] In addition, I offer some reasons why both types of enhancement ought to be affirmed.

I progress as follows. I compare educational and genetic enhancement, showing that Habermas' arguments concerning the relationship between them are implausible. In the conclusion I will refer to the relevance of this insight to the future of education, when the humanities will need to be transformed into metahumanities (Sorgner 2020g, chap 5).

3.3.1 On the relationship between educational and genetic enhancement[24]

Before any intellectual enterprise, the concepts one uses need to be spelled out; for example, it needs to be clarified what is genetic enhancement and what is educational enhancement. Both are difficult to define and have to be described in a broad manner.

In the secondary literature on education there are probably as many definitions as there are experts in the field. The definition I employ is a traditional one, one that is open and not too controversial. Concerning the ethical debate on enhancement, the situation is slightly different, since 'enhancement' as *terminus technicus* is a fairly new philosophical concept. Many ethicists who use the concept leave it undefined in order to avoid the definitional challenges. I will put forward a definition that is closely connected to the concept of eugenics, in order to best evaluate the argument offered by Habermas, who was talking about liberal eugenics and not about 'enhancement'. However, it has become common to use the word and concept 'enhancement'. There seems to be a tendency for bioconservative thinkers to use the term 'eugenics', due to its negative historical connotations, and for bioliberal thinkers to prefer the term 'genetic enhancement', as it is difficult to object to bettering people.

By the concept 'education', I refer to processes that can be described as the general transmission of culture, whereby culture is closely connected to an ideal of the good (for example Eames 1977, 194; Ottaway 1999, 9; Olson 2003, 173; Sorgner 2004). Parents usually play a particularly central role when it comes to the transmission of culture. Obviously, I am not implying that education takes place only if a specific ideal of the good gets transferred. This definition is open to various ideals of the good, so it can be valid for various historical and contemporary settings. I often employ the expression 'educational enhancement' instead of 'education' because, as in other cases of enhancement, the procedure aims at an improvement of the life of the child. An improvement or enhancement is related to a conception of the good, which does not necessarily imply that this conception is a stable one or one that can be described using words.

In the definition of education I used the concept 'parents', and I will employ it again when I specify the concept 'enhancement'. I wish to stress that the concept 'parents', as it is used here, is an open notion that can be specified by talking about biological or cultural parentage. In addition, the concept is limited neither to heterosexual couples nor to heterosexual and homosexual couples: instead, it is conceivable that children can come about by compiling genes from three people of the same sex or by bringing together a sperm cell with an egg cell containing genetic material from two mothers (given a specific mitochondrial disease, this option was legalized in the UK in February 2015). In these cases, all three people involved would be the biological parents. Opening up the concept of parenthood does not render the concept meaningless, and it is needed to differentiate between state-regulated education (or enhancement) and liberal versions.

What about genetic enhancement? It needs to be stressed that eugenics and enhancement are not identical concepts. Eugenics relates specifically to the improvement of genes, whereas enhancement has to do with various types of improvement, whether genetic or otherwise. Eugenics can turn up in a liberal and a state-governed version. The philosophical use of the concept of enhancement in bioethical debates, on the other hand, presupposes a type of liberalism. However, the extensions of the two concepts overlap. I assume that 'liberal eugenics' is a concept that can be subsumed under the concept 'enhancement'. Genetic enhancement by modification and by liberal eugenics have the same extension but different connotations. A fundamental, but often neglected, distinction concerning enhancement is that between *ex post* and *ex ante* enhancement. If a quality X that represents a good is promoted successfully, the outcome is an *ex post* enhancement. If one consciously attempts to promote quality X, then the attempt (or the process employed) can be described as *ex ante* enhancement. In this case, the outcome is uncertain. This is also the case concerning education. On this occasion I am dealing with *ex ante* enhancement (Sorgner 2009).

The decision concerning an enhancement can be made either by oneself (autonomous enhancement) or by one's parents (heteronomous enhancement). Autonomous enhancement is less problematic than heteronomous enhancement. However, primarily in the case of heteronomous genetic enhancement, there is an analogy with educational enhancement. In both of these cases, parents decide what happens to their offspring.

It is never a case of genetic enhancement if the state or a government decides what ought to be done with people, as was done in Germany during the 'Third Reich', since this falls outside the concept of 'enhancement'. As employed in bioethical debates, the notion presupposes a type of liberalism. The notion of 'liberal eugenics' that Habermas employs can be distinguished analogously. The fundamental difference between 'liberal eugenics' and 'enhancement' is that enhancement applies to all types of human qualities. The term 'liberal eugenics' can be employed meaningfully only when genes are altered. Thus, penis enlargement by means of an operation is a type of enhancement, but not an example of liberal eugenics. In this section, I am focusing solely on some problematic cases of genetic enhancement that have the same extension as liberal eugenics, but different connotations. Recall that in this context I am dealing solely with genetic enhancement by modification. I deal with genetic enhancement by selection in the next section (Sorgner 2013).

Many further distinctions concerning enhancement can be made (Sorgner 2006), and I cannot deal with them all. Yet, there is one more that I must mention at least briefly, as it will become relevant later on: the distinction between positive and negative genetic enhancement. Positive genetic enhancement is the conscious attempt to promote good genes. Negative genetic enhancement, on the other hand, refers to an attempt to hinder disadvantageous genes from spreading. The distinction is a problematic one, as the concept 'disadvantageous genes' depends upon a concept of 'disease' that is even more problematic. The more general relationship between positive and negative enhancement is likewise unclear. I will tackle this issue when I turn to Habermas' proposals concerning therapeutic and nontherapeutic uses of enhancement.

The meaning of the concepts genetic and educational enhancement having been discussed, it needs to be spelled out why there could be parallels between these two procedures. In both cases, decisions are being made by parents concerning the development of their child, at a stage where the child cannot yet decide for herself or himself. In the case of genetic enhancement, we are faced with a choice between genetic roulette and genetic enhancement. In the case of educational enhancement, we face the options of a Kaspar Hauser lifestyle or parental guidance. Given these options, it seems most plausible to claim that genetic enhancement and parental guidance usually bring about better results for the offspring than do the alternatives, since the qualities brought about by means of enhancement are based upon parental choices that are normally made on the basis of experience and reflection. Parents usually love their children and want them to have the best possible starting points in life. Of course, parental decisions do not always produce good results. But, as a rule of thumb, parental influence most often leads to better outcomes than those from chance or without any guidance. Parents uphold qualities on the basis of their experience, and having experience in the context of ethical decisions is necessary for making good ethical decisions, as Aristotle remarked about the foundation of prudence.

One difference between the procedures of educational and genetic enhancement could be that education deals with the mind, whereas genetic enhancement relates to the body. However, this point is not raised by Habermas, and it could be answered easily, because (1) it implies a rigid separation between mind and body that is no longer plausible; (2) education also includes physical education; and (3) intelligence and related phenomena that can be enhanced genetically are properties of the mind as well as the body. In addition, (4) I have pointed out that the two procedures are parallel, but not that they are identical.

Habermas has other challenges concerning the differences between genetic and educational enhancement. His main point is that genetic changes are irreversible, whereas educational ones are reversible (Habermas 2001, 90–110). As a consequence, he sees genetic changes as endangering the autonomy of the person in question. He regards the enhancement process as an illegitimate type of instrumentalization of the person and he holds that the consequences of genetic enhancement procedures question the equality of all human beings.[25] However, he regards genetic enhancement as a morally legitimate method in so far as it is employed for clearly therapeutic uses as well as in the case of a prolonged life span as an all-purpose good, because in these cases it is not supposed to attack the autonomy of the person (Habermas 2001, 91). I will deal with these various points one by one. To finish, I will briefly address a final, but invalid, counterargument that has often been raised as a response to one of my presentations on this topic.

Habermas' arguments against genetic enhancement would not be plausible if educational and genetic enhancement were parallel processes, because then the subject status of the enhanced being is touched no more or less in the case of genetic intervention than in the case of educational intervention. The self-perception and understanding of a person who has been genetically enhanced depend upon his evaluation of the enhancement process and his perception of the relationship between education and genetic enhancement. Of course, there is a choice to accept or reject enhancement processes, whether the interventions made are educational or genetic. It is always uncertain whether genetic enhancement will be beneficial. However, it is also an open question whether education will have beneficial consequences in any specific case. However, it is probable in both kinds of cases that the results will turn out better with parental involvement than without it – given that the enhancement methods are reliable.

3.3.2 Irreversibility of genetic enhancement

One claim against a parallel between genetic and educational enhancement is that genetic enhancement is always irreversible (Sorgner 2010b, 4–6). As recent research has shown, however, this claim is implausible, if not outright false.

Let us consider the well-known case of a lesbian couple who were both deaf and chose a deaf sperm donor so as to have a deaf child (Agar 2004, 12–14). Actually, the child can hear a bit in one ear, but this is unimportant for my current purpose. According to the couple, deafness is not a defect; it merely represents being different. The couple were able to realize their wish and managed to have a mostly deaf child. If germ-line gene therapy worked, they could have chosen a non-deaf donor, modified the appropriate genes and brought about a deaf child in this way. Note, however, that if the deafness was one of the inner ear, it would be possible for the person in question to go, later on, to a doctor and ask for a surgical procedure in which he received an implant enabling him to hear. It is already possible to perform such an operation and insert such an implant.

Of course, it may be argued that in that case the genotype is not reversed, but merely the phenotype. This is correct. However, the example illustrates how qualities that came about due to a genetic intervention can be reversible. In this example, they can be changed by means of surgery. But, depending on the type of deafness involved, deaf people may be able to undergo surgery enabling them to hear again.

One could object that the consequences of educational enhancement can be reversed autonomously, whereas in the case of genetic alterations one needs a surgeon or external help to bring about a reversal. This is also incorrect, as I will demonstrate. Meanwhile, it is not true that all consequences of educational enhancement can be reversed. In addition, the availability of somatic gene therapy means that it is even possible to change a person's genetic make-up. One of the most striking examples in this context is siRNA therapy, by means of which genes can be silenced. In the following paragraph I give a summary of what siRNA therapy has achieved so far.

In 2002 the journal *Science* referred to RNAi as 'Technology of the Year', and McCaffrey et al published a paper in the journal *Nature* in which they specified that siRNA functions in mice and rats (McCaffrey et al 2002, 38–39). Evidence that siRNA therapy can be used successfully in animals was published by Song et al in 2003. By means of this type of therapy (RNA interference targeting Fas), mice can receive protection from fulminant hepatitis (Song et al 2003, 347–351). A year later, it was shown that genes at a transcriptional level can be silenced by means of siRNA (Morris et al 2004, 1289–1292). Due to the enormous potential of siRNA, Andrew Fire and Craig Mello were awarded the Nobel Prize in medicine for discovering the RNAi mechanism in 2006.

Given the empirical data concerning siRNA, it is plausible to claim that theoretically the following process is possible, and, hence, that genetic states are not necessarily immutable: (1) an embryo with brown eyes can be selected by means of PGD; (2) the adult does not like his eye colour; (3) he asks physicians to provide siRNA therapy to change the gene related to his eye colour; (4) the alteration brings about an eye-colour change.

Another option would be available, if germ-line gene therapy worked, which it does not in a reliable manner so far.[26] In that case, we could change a gene using germ-line gene therapy to bring about characteristic X. Imagine that this characteristic is disapproved of by the later adult. Hence, he decides to undergo genome editing in order to silence the altered gene again. Such a procedure is theoretically possible. However, we do not have to use fictional examples to show that alterations brought about by genetic enhancement are reversible; we can, instead, simply look at some developments in the field of gene modification.

A 23-year-old British male, Robert Johnson, suffered from Leber's congenital amaurosis, which is an inherited blinding disease. Early in 2007 he had surgery at Moorfields Eye Hospital and University College London's Institute of Ophthalmology, which represented the world's first gene therapy trial for an inherited retinal disease. In April 2008 the *New England Journal of Medicine* published the results of this operation, which revealed its success, as the patient had gained a modest increase in vision with no apparent side-effects (Maguire et al 2008).

This case involved a therapeutic use of genetic modification. However, genes that can be altered for therapeutic purposes could also be altered for non-therapeutic ends (if we wish to uphold the problematic distinction between therapeutic and non-therapeutic ends). The examples mentioned here clearly show that qualities brought about by means of genetic enhancement do not have to be irreversible. As we will see, the parallels between genetic and educational enhancement go even further.

3.3.3 Reversibility of educational enhancement

According to Habermas, character traits brought about by educational means are reversible (Habermas 2001, 110–111). Because of this assumption, he rejects the idea that educational and genetic enhancements are parallel processes. Aristotle disagrees, and he is right in doing so. According to Aristotle, a *hexis*, a basic stable attitude, becomes established by means of repetition.[27] If you continually act in a brave manner, you become brave. By playing a guitar, you turn into a guitar player. By acting with moderation, you become moderate. Aristotle states that by repeating a certain type of action, you establish the type in your character: you form a basic stable attitude, a *hexis*. In the *Categories*, he makes clear that the *hexis* is extremely stable.[28] In the *Nichomachean Ethics*, he goes even further and claims that once one has established a basic stable attitude, it is impossible to get rid of it again.[29] Buddensiek has interpreted this passage correctly by pointing out that once a *hexis*, a basic stable attitude, is formed or is established, it is, according to Aristotle, an irreversible part of the individual's character (Buddensiek 2002, 190).

Aristotle's position receives support from Freud, who put forward the following claim: 'It follows from what I have said that the neuroses can be completely prevented but are completely incurable' (Malcolm 1984, 24). Angstneurosen were supposed to be particularly striking examples (Rabelhofer 2006, 38).

Much time has passed since Freud, and further research has taken place. In publications concerning psychiatric and psychotherapeutic findings, however, it remains clear that psychological diseases can be incurable (Beese 2004, 20). Of course, psychiatric disorders are not intentionally brought about by educational means. However, much empirical research has been done in the field of psychiatric illnesses and their origin in early childhood. The robust finding that irreversible illnesses can come about from events or actions in childhood entails that irreversible outcomes can happen by means of proper educational measures.

Medical research has shown, and most physicians agree, that post traumatic stress disorders can not only become chronic but also lead to permanent personality disturbances (Rentrop et al 2009, 373). They come about as a result of exceptional events that represent an enormous burden and change within someone's life. Obsessional neuroses are another such case. According to the latest numbers, only 10% to 15% of patients get cured, and in most cases the problem turns into a chronic disease (Rentrop et al 2009, 368). Yet another example is provided by the borderline syndrome, which is a type of personality disorder. It can be related to events or actions that have taken place in early childhood, such as violence or child abuse. In most cases, this appears as chronic disease (Rentrop et al 2009, 458).

Given these examples, it is clear that actions and events during one's lifetime can produce permanent and irreversible states. In the psychiatric examples, the outcome is a disadvantage to the person in question. In the case of an Aristotelian *hexis*, by contrast, it can be advantageous if the person establishes a virtue in this manner.

To provide further intuitive support for the position that qualities established by educational enhancement can be irreversible, simply think about learning to ride a bike, tie one's shoe laces, play the piano, or speak one's mother tongue. Children are educated for years and years to undertake these tasks. Even when one moves into a different country, or if one does not ride a bike for many years, it can be difficult, if not impossible, to remove the acquired ability. Hence, it is very plausible that educational enhancement can have irreversible consequences, and Habermas is wrong again. Genetic enhancement can have consequences that are reversible, and educational enhancement can have consequences that are irreversible. Given these insights, the parallel between genetic and educational enhancement obtains additional support. However, I will consider some further points that Habermas raises.

3.3.4 Autonomy

To support his main point of critique – his denial of a parallel between educational and genetic enhancement – Habermas raises many further questions. According to him, genetic (but not educational) enhancement limits the potential for an autonomous way of life (Habermas 2001, 45). To support this claim, he explains that there exists a clear distinction between something that has grown and something that was made in the lifeworld (Habermas 2001, 83). Only human beings who have solely grown are supposed to have their full autonomy.

The distinction between what has grown and what was made is problematic. It seems highly implausible to hold that human beings who grow up are solely growing up. Human beings are in a permanent interaction with their environment and their culture and they are also influenced by whatever they get to eat and drink. In addition, could one not argue that we are already making human beings? This happens if a woman goes to a sperm bank and asks for the sperm of a Nobel Prize winner, which can be done in the US, although sperm from good-looking, intelligent, and athletic Ivy League students has proved to be more popular (Agar 2004, 1–2).

In another sense, we are already making human beings whenever we choose partners with whom we can have children. When we decide to have unprotected sex at a certain time, we are potentially making human beings. To hold that only human beings who are genetically enhanced are being made is too simple and rigid a position to be plausible, particularly if we consider the consequences of educational enhancement in more detail. For example, a child who grows up in an extremely religious environment and so receives a religious education can be indoctrinated irreversibly. In such a case, we might claim that the child was made, as the child acquired characteristics that he or she cannot remove. Habermas might not agree that this is possible, since he might reply that the grown autonomous subject decides which educational means he or she does or does not accept. This would accord with his emphasis on the rationally motivated affirmations of an independent subject (Habermas 2001, 99). At this point, however, it becomes clear that he is clinging to an anthropology within which human beings have a special status, since they and only they are supposed to be rational and independent subjects. Although Habermas claims explicitly that he is in favour of a 'soft naturalism', he uses the concept of a special subject that is beyond any empirical analysis (Habermas 2004, 876–877). He puts forward a view of human beings that is extremely implausible after Darwin, Nietzsche, and Freud and after post- and transhumanism.

Habermas draws further inferences. He holds that genetic enhancement might cause a break within humanity: it might divide us into human beings who are grown and autonomous, and human beings who were heteronomously made and are therefore less autonomous. His description implies that the less autonomous ones are somehow inferior (compare the 1997 film *Gattaca*, which depicts a situation in which genetically selected humans regard themselves as superior). No matter what the consequences may be, Habermas holds that genetic enhancement touches a question concerning the identity of a species (Habermas 2001, 45). In a way, he is correct, since genetic enhancement could, in principle, bring a new species into existence. Some transhumanists refer to human beings who develop the potential of or who are on the way to becoming members of a new species as 'transhumans' and to the members of a new species as 'posthumans'. However, we cannot exclude the possibility that the same result could come about by educational enhancement. Nietzsche held that, by means of educational enhancement, we can establish preconditions for the next evolutionary step to occur, so that a new species of *overhumans* can come into existence (Sorgner 2009). Hence, the identity of our species could be altered not only by genetic enhancement but also by educational enhancement.

Habermas goes even further in his critique. He correctly holds that people have the right to an open future, but then claims that genetic enhancement limits the life plans of the enhanced people, as their freedom of choice will have been limited (Habermas 2001, 105). To be autonomous, human beings must be the sole authors of their way of life (Habermas 2001, 109). Habermas' claim is simply false, however, as the freedom of choice of a genetically enhanced human being is not limited but is merely altered as compared to the non-enhanced person.[30] Every human being has a genetic make-up. The question is who decides upon and brings about the genetic make-up. In the one case it is chance, and in the other case it is a parental decision. The parents do not limit the decisions of their child, but merely alter the preconditions. A child who is not genetically enhanced also has a genetic make-up that determines some of her or his strengths and weaknesses.

3.3.5 Instrumentalization

Habermas raises still another issue. He claims that in using genetic enhancement parents will instrumentalize their children, as a child cannot object to what happens to him. Yet, instrumentalization takes place whenever person X uses person Y merely as a means to an end. For comparison, consider some problematic cases of educational enhancement, for example little girls living the dreams of their mothers (perhaps becoming a model, being a nun, or being absolutely spoiled with luxury goods). Habermas explains that it might make us sick to imagine that our nature was instrumentally altered before birth (Habermas 2001, 94), as such a procedure might have significant consequences upon our self-understanding. But this does not have to be the case if we understand educational and genetic enhancement as parallel events. If they are understood as parallel, then the consequences of the one would not be better or worse than those of the other.

In addition, we can doubt that it ought to be prohibited to use a person solely as a means to further ends. Hoerster has presented a good example against the absolute validity of that prohibition, and has suggested plausibly that we can distinguish between morally legitimate and morally illegitimate types of instrumentalization (Hoerster 2002, 15). In Chapter 4, section 3, I will provide further reasons why mere instrumentalization needs to be seen as an implausible concept. Still, one further thought needs to be added: if we prohibit genetic enhancement because human beings are instrumentalized during the process, then educational enhancement should also be forbidden. However, we can reject the central accusations that during the process of genetic enhancement parents are merely instrumentalizing their child. The accusation is false: it is (usually) not a case of a mere instrumentalization, since the parents also love and respect the child and this influences the process of genetic enhancement. The parents might partly instrumentalize the child, but that is not, in itself, morally wrong. By way of comparison, an employer partly instrumentalizes her employee, but this does not entail treating the employee immorally. In short, there are many plausible reasons that explain why Habermas' position – his accusation that a child is treated immorally during the process of genetic enhancement – is highly implausible.

Although there are many reasons for rejecting Habermas' position concerning instrumentalization, the most important one is more fundamental than anything discussed up to this point. The moral objection to using persons solely as a means presupposes a radically dualistic ontology that is highly dubitable. To explain this in more detail, I will elaborate on this issue in Chapter 4, section 3 (see Sorgner 2013d).

3.3.6 Equality

A related but distinct issue is that of equality. Habermas claims that genetic enhancement – but supposedly not educational enhancement – destroys symmetrical relationships among free and equal people (Habermas 2001, 45). He supports this by reference to his distinction between the grown and the made, along with his reflections on autonomy. If the genes of someone, person X, are altered irreversibly by someone else, person Y, but X cannot bring about the same type of changes in Y, then this creates an asymmetrical relationship that will, supposedly, destroy the relationship of equality. As genetic changes do not have to be immutable, however, this is a false concern. Furthermore, even if such an intervention created an asymmetrical relationship between person X and person Y, this need not have any effect upon equality as a normative ideal.

In addition, ordinary kinds of education can bring about states that are irreversible. Hence, processes are currently being used that create asymmetrical relationships without any grave moral problems. Simply being a parent necessitates being in an asymmetrical relationship with one's children. It does not follow, however, that equality as a normative ideal will have to be abandoned just because some human beings are related in an asymmetrical manner.

We might wonder what type of equality Habermas has in mind. If equality can exist only between identical entities and we assume the strongest version of the Leibnizian concept of identity, then we have to conclude that no equality can exist in the empirical world between two distinct objects. If Habermas has a type of normative equality in mind, then I see no compelling reason why it would have to be given up if some human beings were genetically enhanced.

3.3.7 Therapy and enhancement

In contrast to his negative remarks about genetic enhancement, Habermas does accept that gene therapy can be morally legitimate in at least some cases (Habermas 2001, 109), even though it has the consequences discussed earlier. This seems to be a self-contradictory, or at least highly problematic, position. Gene therapy is morally legitimate, according to Habermas, as it does not undermine the autonomy of the subject. But this judgement does not apply to genetic enhancement, since it technically alters human nature (Habermas 2001, 192). This position seems problematic, if he regards it as dangerous that the limits of our species are altered by means of genetic enhancement per se. We could wonder if this worry does not already apply in the case of gene therapy.

Habermas is sceptical concerning most all-purpose goods (intelligence, humour, patience, and so on), which are goods that support all conceptions of a good life.[31] Consequently, he is critical of genetic enhancement, as he does not think that we can have a catalogue of goods that are actually beneficial for all human beings, but he thinks that such a catalogue would be needed for the process to be a morally legitimate one. Despite all this, he upholds the values of health and a longer life (Habermas 2001, 91), and regards parents' care for these qualities as corresponding with the autonomy of their child (Habermas 2001, 48). Many critical questions must be raised concerning this judgement.

Firstly, I wonder why genetic enhancement aimed at promoting the life span is morally illegitimate. Habermas has clearly said that it is legitimate for the parents to make decisions for the child that promote the child's life span and that such decisions do not interfere with the child's autonomy. He claims the same concerning decisions about the child's health, which suggests that he does not regard gene therapy as morally objectionable. If this is correct, then he should not object to genetic enhancement that aims for a longer life span of the child.[32]

Secondly, it needs to be stressed that a gene diagnosis that is a prerequisite to genetic enhancement and gene therapy already includes an alteration of the genes (Koechy 2006, 75–77). Hence, gene therapy, of which he approves (in some cases, at least), presupposes a process, the alteration of genes, of which he disapproves. This looks like an unstable position.

Thirdly, I need to stress that there is no clear-cut distinction between therapy and enhancement. The concept of therapy presupposes the concept of disease. However, the definition of the concept 'disease' is highly problematic: (1) if we wish to give an objective definition of 'disease', then we need to have a natural understanding of what human beings are. As disease is a normative concept, this creates problems, because we would get statements like: a natural being is so and so tall, has capacities A, B, and C and has such-and-such a sexual orientation; (2) subjective definitions of the concept 'disease' include many problems. Someone is ill, if he feels bad. But this definitely does not have to be the case with cancer. One can have a malignant tumour without feeling it in the beginning; (3) the concept 'disease' changes over time. This becomes particularly clear if we consider the history of the concept, particularly concerning psychological diseases. Experience shows that the concept can be manipulated to further the interests of political leaders. Given all these reflections, I conclude that Habermas' position concerning disease is highly problematic. This insight has further support due to the fact that Habermas' notion of therapy on the one hand seems to include processes which represent prime examples of enhancement procedures (prolonging the health span or preventive measures) (Habermas 2001, 91), but on the other hand seems to be limited solely to extreme maladies (Habermas 2001, 109).

As Habermas does not provide us with any clear definition of what a disease is, we can use a concept of 'disease' which implies that procedures that many would refer to as genetic enhancement are actually merely a type of therapy. In so far as Habermas regards gene therapy as morally defensible, he would have to approve the actions in question.

3.3.8 Educational enhancement is necessary but genetic enhancement is not

Finally, I wish to address a counterargument against my thesis concerning the parallel between genetic and educational enhancement. Though Habermas does not put this forward, it has been raised against my position at a number of presentations that I have given on this topic. It does rely on a point mentioned by Habermas – that all newborns are in need of help (Habermas 2004, 884). Accordingly, one could argue the following.

1. Newborns need human support to survive.
2. All human support is a type of education.
3. Genetic enhancement, on the other hand, is not necessary for survival.
C. Therefore, genetic and educational enhancements are not parallel processes.

On this argument, genetic enhancement is distinguishable from morally supportable forms of educational intervention in not being necessary for the survival of the child.

One can reply in various ways. However, the most important reply is associated with the latest epigenetic research, which reveals the intimate interconnectedness between educational and genetic enhancement. Nietzsche put forward education as a means to bring about the posthuman. Given epigenetics, he thereby implicitly also affirms genetic enhancement.

Still, one could wonder: can education bring about changes that have an influence on the potential offspring of the person who is educated? As inheritance depends upon genes, and genes do not get altered by means of education, we used to believe that education cannot be relevant for the process of evolution. Hence, Lamarckism, the heritability of acquired characteristics, has not been very fashionable for the same period of time. However, in recent decades doubts have been raised, based upon research in epigenetics. Together with Japlonka and Lamb, I can stress that 'the study of epigenetics and epigenetic inheritance systems (EISs) is young and hard evidence is sparse, but there are some very telling indications that it may be very important' (Japlonka and Lamb 2005, 248).

Besides the genetic code, the epigenetic code, too, is supposed to be relevant for creating phenotypes and it can be altered by means of environmental influences. The epigenetic inheritance systems belong to three supragenetic inheritance systems that Japlonka and Lamb distinguish. These authors stress that 'through the supragenetic inheritance systems, complex organisms can pass on some acquired characteristics. So Lamarckian evolution is certainly possible for them.'[33]

Given recent work in this field it is likely that stress,[34] education,[35] drugs, medicine, or diet can bring about epigenetic alterations that, again, can be responsible for an alteration of cell structures (Levenson/Sweatt 2005, 108–118) and for the activation or silencing of genes (Levenson/Sweatt 2005, 108–118). In some cases, the possibility cannot be excluded that such alterations might lead to an enhanced version of evolution. Japlonka and Lamb stress the following: 'The point is that epigenetic variants exist and are known to show typical Mendelian patterns of inheritance. They therefore need to be studied. If there is heredity in the epigenetic dimension, then there is evolution, too' (Japlonka/Lamb 2005, 359).

They also point out that 'the transfer of epigenetic information from one generation to the next has been found and that in theory it can lead to evolutionary change' (Japlonka/Lamb 2005, 153). Their reason for holding this position is partly that 'new epigenetic marks might be induced in both somatic and germ-line cells' (Japlonka/Lamb 2005, 145).

A 'mother's diet' can also bring about such alterations, according to Japlonka and Lamb (Japlonka/Lamb 2005, 145); hence the same potential as the interventions previously discussed (genetic enhancement and education) logically applies to the next method of bringing about a posthuman, that is, non-genetic enhancement by means of drugs, medicine, or diets. As has become clear already, such measures can lead to an enhanced version of evolution, given recent research in the field of epigenetics. Given these insights, it is clear that educational and genetic enhancements are processes that do not exist independently of one another. If one process is necessary, this also applies to the other. Education can be a genetic enhancement via epigenetic processes.

3.3.9 Conclusion

Given the earlier analysis, I conclude that Habermas is wrong when he denies that educational and genetic enhancements are parallel events. In addition, I have mentioned some reasons why educational enhancement in most cases is better than undergoing a Kaspar Hauser type of development. We can also conclude that genetic enhancement ought to be affirmed – analogously to our affirmation of educational enhancement – with no need to settle distinguishable questions, such as the moral status of the embryo or which conception of the good to employ when making decisions about education and enhancement.

In order to reach a clearer understanding of which types of genetic enhancement should (and should not) be undertaken, we would need to consider the moral status of the embryo and which conceptions of the good ought to apply on a political level. For current purposes, I stress that negative freedom is a precious achievement. During and since the Enlightenment, we freed ourselves from the paternalistic oppression of religious and aristocratic leaders. Thereby, we established the right to live according to our respective conceptions of the good, as long as this does not interfere with the rights of someone else. Consequently, I suggest that *in dubio pro libertate* is an adequate principle for a democracy. If there is a conflict between several groups beyond a certain significant size, then the opinion ought to be legalized in favour of more freedom. Hence, the state should refrain from making demands based upon metaphysical and religious prejudices.

If we apply this norm to questions concerning the status of the embryo, it follows that this is unclear metaphysically. On this occasion, I cannot provide an in-depth discussion of the moral status of the embryo, as I will do so in Chapter 4, section 3, but I will make a few brief comments. There is a group that attributes embryos as well as adult human beings the moral status of being persons. However, there are other large groups that regard an embryo as a collection of cells that ought to be given special consideration because of their implicit potential, but not the same rights as a human being. A liberal state would have to allow both groups to live in accordance with their principles, as long as they do not interfere with the rights of others. Here we can identify another parallel between genetic and educational enhancement: in both cases, parents make decisions about the lives of other human beings that do not yet have all human rights.

If it is accepted that genetic enhancement by modification and education are structurally analogous processes, this insight needs to be considered when thinking about the future of parental education under the altered cultural conditions of recent decades. Emerging technologies, human–machine interfaces, and new scientific insights are changing our way of understanding the world. Transformative sciences and technologies seem to dominate the way the world works, and to carry out whatever projects are feasible. The humanities seem to be out of step amid rapid scientific and technological developments, and consequently their relevance seems to diminish.

We are living in an age of bioengineering and digital technologies. However, the most fundamental and urgent questions concerning ethical, ontological, legal, political, social, and cultural issues cannot be addressed appropriately solely by means of the natural sciences and engineering, as this task lies outside the scope of their expertise. To approach all the latest questions in a thoughtful and comprehensive way, we need informed intellectual reflection, insights concerning our place in cultural history, and an awareness of the great plurality of philosophical, ethical, and religious positions that have been dominant in human history. At the same time, it is an open question whether specialists, experts, and scholars from the humanities in its traditional form possess all the necessary skills. I doubt that they do, and I also doubt that they start from appropriate premises, because they assume that solely human beings are categorically ontologically separate from all other natural beings.

The concept of 'humanities' in its traditional sense is connected to the term *humanitas* and to the *artes liberales*. Both concepts are closely related to ancient times about 100 BCE: to Varro, Cicero, and others. Since then, these concepts have been connected to the affirmation of categorical dualities. Thanks to Kant's account of issues relevant for the humanities, categorical dualities are still associated with the central concepts in question: for example, mind–matter, culture–nature, genes–environment. Yet, it is this aspect of the humanities that has been challenged by the latest scientific insights and discoveries, and by technological developments. Epigenetics, posthumanism, transhumanism, embodied theories of mind, and further scientific research all cast doubt on the affirmation of categorical dualities, and have inspired attempts to move beyond this way of conceptualizing the world. This has severe consequences for many different fields of enquiry, even for ethical, legal, and political issues: for example, questions relating to autonomy and the supposed moral prohibition of treating a person solely as a means. To consider all of these implications with due seriousness, we will have to move away from the traditional humanities and towards an approach that we can refer to as 'metahumanities', that is, the humanities need to be become more inclusive concerning topics dealt within 'Posthuman Studies'. This move has particularly relevant implications for the question of education in an age of transformative sciences and emerging technologies. I suggest that the following three insights need to be considered as implications of founding education upon the emerging metahumanities, whereby the initial two suggestions are relevant for parental education and the final suggestion is one that needs to be considered in schools and universities.

First, genetic enhancement by modification and education are structurally analogous processes. In this chapter, I have argued in detail that this is the case. Second, gene analysis will become a prerequisite for a well-informed education. Bioprivacy and big gene data will be the keywords in this context, and I expect these keywords to lead to intense future discussions, as well as enormous revisions in the field, of the future of occupations and insurance companies as well as education. Third, the categorical distinction between genetic and environmental influences will dissolve, and the relationships between these influences will form an additional school subject within the metahumanities. It will involve consideration of all the following: bioarts; evolutionary epistemology, aesthetics, ethics, economics, and so on; embodied theories of the mind; epigenetics; new types of spirituality and mysticism; non-dualist accounts of rights and dignity; revised concepts of the family; naturalistic conceptions of the good life; the relevance of cultural history with respect to norms, values, and so on; and difficult questions relating to non-anthropocentric ontologies and the avoidance of speciesism.

In this section, I have provided a number of detailed reasons for the first suggestion. The parallel between genetic enhancement by modification and education is one insight that must be considered when discussing the future of education. In other words: we can expect that genetic enhancement by modification will be a central issue for the future of parental education.

3.4 Gene selection[36]

Julian Savulescu claims that human beings have a 'moral obligation to create children with the best chance of the best life' (Savulescu/Kahane 2009, 274), because he regards the principle of procreative beneficence, abbreviated as PB, as morally right in the context of choosing fertilized eggs after IVF and preimplantation genetic screening (PGS), which is a different gene technology than the one I dealt with before, that is, gene modification, genome editing, CRISPR-Cas9. According to this principle 'couples who decide to have a child have significant moral reason to select the child who, given his or her genetic endowment, can be expected to enjoy the most well-being' (Savulescu/Kahane 2009, 274). PB has been criticized by several scholars since it has been presented (De Melo-Martin 2004; Birch 2005; Herissone-Kelly 2006; Parker 2007; Sparrow, 2007; Sparrow, 2011). In most cases the critics did not consider adequately their position or implied counterarguments which are irrelevant for their line of thought. However, I think the scholars were right in rejecting Savulescu's principle of PB. From my perspective, it is morally appropriate, if one deals with challenges related to the creation of children, to take the principle of procreative autonomy (PA) into account, which is a principle Savulescu rejects. It is an issue of reproductive freedom.

In sub-section 1, I will briefly describe the procedures with which Savulescu is concerned, and I will present Savulescu's argument in favour of the principle of PB. In sub-section 2, I will show that the principle is inconsistent and that it violently attacks human beings who disagree with it, which is the reason why I regard it as an immoral principle. In the conclusion, I will put forward some reasons for regarding the principle of autonomy (PA) as morally more plausible than PB, regarding the questions Savulescu deals with.

3.4.1 Creating children and the principle of PB

Earlier, Savulescu talks about creating children. Creating children goes beyond merely bringing new children into existence but implies actually doing something to influence the genetic make-up of one's offspring. This can be done in various ways. Two methods are particularly prominent when technologies of genetic enhancement are being discussed: (1) creating a child by selecting a fertilized egg after IVF and PGD (genetic enhancement by selection); (2) creating a child by actively altering a gene of an already given genetic make-up, which could be done by CRISPR-Cas9 or by means of transduction, whereby one introduces a modified virus into the cell which gradually changes a certain gene in all of someone's somatic cells (genetic enhancement by modification).

When Savulescu talks about the principle of PB, he merely has the first option in mind, genetic enhancement by selection, as he claims the following: when we have the reliable and safe option of choosing fertilized eggs after IVF and PGD, then we have the moral duty to choose the entity with the best chances of the best life (Savulescu/Kahane 2009, 274).

Here it can be asked whether the probability or the quality of life ought to be considered most, but this is not a crucial concern of mine. In the 2009 version of the principle of PB, which Savulescu formulated together with Guy Kahane, it is clear that his principle applies only to the method of selecting a child:

> If couples (or single reproducers) have decided to have a child, and selection is possible, then they have a significant moral reason to select the child, or the possible children they could have, whose life can be

expected, in the light of the relevant available information, to go best or at least not worse than any of the others. (Savulescu/Kahane 2009, 274)

Savulescu does not claim that we have the moral obligation to choose the best child among the 50 or 60 fertilized eggs which were created during one IVF process, but he has in mind that we have duty to choose the best child among the totality of children which can come about during all potential processes of IVF. If no suitable entity is there this time, then we ought to try one or several more times (Savulescu 2001, 417).

Given his utilitarian approach (not in its act-utilitarian version), he calculates as follows determine what is morally appropriate. He compares two events or situations, namely the overall utility before a family has created a child and after it has taken place. According to his moral philosophy, the act is morally right which brings about the most overall utility. Given his reflections, this is the case when a family has a child with the best chance of the best life, because then the overall utility will have been maximized. The example which he presents to support his line of reasoning is the following:

Imagine now you are invited to play the Wheel of Fortune. A giant wheel exists with marks on it from 0±$1000000, in $100 increments. The wheel is spun in a secret room. It stops randomly on an amount. That amount is put into Box A. The wheel is spun again. The amount which comes up is put into Box B. You can choose Box A or B. You are also told that, in addition to the sum already put in the boxes, if you choose B, a dice will be thrown and you will lose $100 if it comes up 6. Which box should you choose? The rational answer is box A. Choosing genes for non-disease states are like playing the Wheel of Fortune. (Savulescu 2001, 414)

Savulescu's example works as a thought experiment. If you wish to maximize your money, it is rational to choose box A. He implies that there is an analogy between his boxes and the genetic make-up of the aforementioned case and he assumes that in any given situation we have one state of genetic make-up which clearly has the best overall utility.

3.4.2 Counter-arguments against the principle of PB

In the following sections, I will first present some general counterarguments and then some more specific and crucial ones to show that his principle is inconsistent and immoral, as it implies a cruelty and violence against minority groups which do not agree with his ethical theory.

3.4.2.1 General concerns

Let me return to Savulescu's analogy between boxes and disease states. If you wish to make money, choosing box A is the rational decision. However, his apparently plausible analogy between boxes and genetic make-up does not work, because when we deal with genes and disease states the question of what is best is not answered that easily. Let us assume that IVF has taken place and afterwards some fertilized eggs underwent PGD. Consequently, we know that the fertilized egg *A* has characteristics *a1*, *a2*, and *a3*, egg *B* has *b1*, *b2*, and *b3*, and egg *C* has *c1*, *c2*, and *c3*.

We know that entity A will have an above average intelligence, have an excellent memory, and probably be homosexual; entity B will be physiologically strong, promises to be exceptionally healthy, and will have an average height; and entity C will be exceptionally intelligent, extremely aggressive, and have a prolonged life expectancy.

Which entity is supposed to be the best one? Is it obvious to all of us what counts as a negative trait? In addition, the line of argument becomes more complicated, because Savulescu stresses that the principle does not apply to several choices at only one time, but potential future children also need to be taken into consideration, which becomes clear when he talks about Parfit's rubella example and the case of the nuclear accident (Savulescu 2001, 417–418).

Let us assume that a family wishes to have one child. Given Savulescu's PB, they have the moral obligation to choose the child with the best chance of the best life. Hence, parents also need to consider the following. At time *t1*, they can choose between the afore-mentioned entities A, B, and C. None of them is exceptionally intelligent, has an outstanding memory, possesses a strong health, and has a prolonged life expectancy, which nonetheless would be the desired combination of the couple in question. Consequently, they also wish to consider entities D, E, and F at time *t2*, which is a couple of months later, and check whether their genetic make-ups are more promising. However, it is impossible to compare entities A, B, and C with entities D, E, and F because the qualities of D, E, and F cannot be known. The option of making a comparison can only be provided if we freeze fertilized egg cells A, B, and C, and compare them to D, E, and F once they are available for making a comparison.

As you can never know whether A, B, or C is better than a later D, E, or F, parents will always be obliged to undergo a new IVF, given PB, because it can always be the case that the later entities will be better than the ones currently available. When is it the case that a couple is morally allowed to stop searching for a better-fertilized egg, according to PB? A moral theory which demands choosing the best child in such circumstances is not a helpful theory, because it does not have a serious practicable applicability.

The principle of PB does not apply only to IVFs. Understood in a wider sense, without its being restricted only to selection procedures, it is open to various processes of fertilization. The principle can also imply that couples are morally obliged to use the method of an IVF instead of relying on the 'natural' method for reproduction. In the case of sexual intercourse, the genetic make-up comes about by chance, whereas in the case of IVF, parents can choose the best child. As the probability that the best child will come into existence is highest when parents have the option of choosing their child, the principle implies that we have both a moral duty to avoid sexual intercourse as a method for procreation as well as a duty to use a method of contraception to make sure that sexual intercourse will not be one's method of reproduction. However, I do not think that this is a crucial counterargument against Savulescu's principle, because he stresses that his principle is a *pro tanto* obligation, which implies that it can be overruled by other insights, for example maybe the moral value of having a child by means of sexual intercourse.

3.4.2.2 Inconsistencies

In this section I will present some reasons for holding that Savulescu's web of thoughts associated with his PB is inconsistent. This is the case, because he refers to at least two incompatible standards of goodness within his argument. When he describes his principle of PB he refers to specific qualities which are supposed to be associated with a good life. Thereby, he mentions being healthy, strong, intelligent, long lived, and having a strong memory. However, when he replies to counter-arguments against his principle he alters his concept of goodness and suddenly allows and focuses on other factors such as external circumstances and social settings, which becomes particularly clear in the latest reformulation of his principle, which he published in the article 'The Moral Obligation to Create Children with the Best Chance of the Best Life' (Savulescu/Kahane 2009).

In one sentence, Savulescu claims that PB is neutral to central philosophical issues concerning the good life. He explicitly says that 'PB is neutral with respect to such philosophical disputes about the nature of the good life' (Savulescu/Kahane, 2009, 279). In the paragraph before this sentence, he talks about various philosophical theories of the good life. However, in the next sentence he affirms the following:

> But although there is this philosophical disagreement, there is considerable consensus about the particular traits or states that make life better or worse, a consensus that would rule out many procreative choices as grossly unreasonable [...] PB doesn't rely on some special

and controversial conception of well-being. All it asks us is to apply in our procreative decisions the same concepts we already employ in everyday situations. (Savulescu/Kahane 2009, 279)

Here, he makes clear that PB is not neutral to philosophical theories of the good life, but that he upholds a common-sense type of approach to the good life. He is also confident to know that 'there are plenty of cases where we can rank the goodness of lives. We do so in numerous moral decisions in everyday life' (Savulescu/Kahane 2009, 279). According to Savulescu, we know what a good life is, and what we regard as a good life actually is what is needed for a good life. Yet, he does not specify further who is the 'we' to whom he is referring. Does he refer to the majority of people in Western countries, fellow intellectuals at the University of Oxford, or all strong interest groups in Western countries, like the US country folk?[37]

In a further paragraph, he slightly alters his concept of the good life again, because he puts forward that it is supposed to be clear that there is such a thing not only as the good life, but even as the best life, as he claims: 'a common objection to the PB is that there is no such thing as a better or best life. It is hard to defend such a claim' (Savulescu/Kahane 2009, 278).

The validity of there being a good life can imply that there are some general principles which are valid for all human beings, which is already an extremely strong claim. However, Savulescu in this argument moves beyond the affirmation of the good life and towards the belief that there is actually a best life. Plato held such a position which he managed to explain by reference to his theory of forms. However, for a libertarian, utilitarian philosopher such as Savulescu to uphold this claim is quite a daring position. I would be keen to know how he can manage to explain this theoretical foundation of the PB.

Savulescu does not only make general remarks concerning the question of the good, but actually puts forward a list of qualities what we regard as good and bad, according to his perspective. In section 4.2.2, I present a detailed summary of his reflections on the good life.

Yet, I am not able to order his various utterances such that they fit together consistently. It rather seems to me the case that he upholds various mutually exclusive theories of the good which he uses in order to have plausible replies to his critics' concerns.

3.4.2.3 PB as a violent and hence an immoral principle

According my point of view, PB is not only inconsistent but is also an immoral principle because it acts violently against individuals and interest groups who do not agree with PB and the associated theory of the good, or, should I say, the corresponding theories of the good. Savulescu himself is aware that PB is a much stronger theory than the theories of the good which most liberal ethicists have proposed in recent years, and he regards it as morally appropriate that his is such, as he clearly holds the following view: 'Although PB and the procreative principles we have considered here bear little resemblance to the collectivist, coercive, and often racist projects of 20th century eugenics, most supporters of genetic selection have tended to proceed gingerly, defending views that are unnecessarily weak' (Savulescu/Kahane 2009, 282).

I think that Savulescu is right in stressing the difference between his PB and procedures of state-governed eugenics during the German 'Third Reich', even though some commentators have criticized him for proposing a new type of eugenics (for example Sparrow, 2011). However, you do not have to propose fascist views for holding an immoral view. He even regards PB as a moral principle: 'PB is a moral principle. It states what would be morally right or wrong for reproducers to do' (Savulescu/Kahane 2009, 279).

Let me put forward some reasons for regarding PB as immoral. Firstly, I regard it as immoral because it is violent against people who do not subscribe to it. Secondly, I see it as immoral because it implicitly contains immoral duties.

How far can it be said that PB acts violently against people who do not subscribe to it? PB acts violently because it implies that parents who do not subscribe to the theory of the good life as Savulescu proposes it act immorally. The principle demands to tell these parents that they ought not to have acted the way they did, and that thereby they have acted falsely. In this way, the principle intrudes paternalistically in the lives of other people and acts violently against their concept of the good life.

Parents might decide not to choose the fertilized egg which has the greatest memory, because they know that the child will grow up in a poor family and during times of war, and they think that it will most probably be good for the child if the child does not have to remember all the bad things which are bound to take place, given such problematic circumstances.

PB implies that if parents prefer someone with a weaker memory to someone with a stronger memory, they are acting immorally. I do not think that parents have acted immorally in these circumstances but, rather, that they have made a decision concerning the good life which is understandable, given the circumstances they live in.

Savulescu might reply that PB allows parents to consider the circumstances in the process of evaluating which qualities increase the child's probability of living a good life, as he dealt with this issue analogously in the earlier case of disability. As mentioned before, he also made clear that this position is valid only for the case of disabilities, and, given the other phrases cited earlier, he holds that having a stronger memory is better than having a weaker one. If, in the given context, he claims that the social circumstances ought to be considered too, it becomes clear that he holds mutually exclusive concepts of the good life. If he claims that the choice of the parents in question was immoral, he upholds an immoral position because he acts violently against the view of the good of the parents and acting violently in this manner is an immoral act.

Besides PB being violent, it also implies immoral duties. Savulescu fails to see the impact of a distinction to which he himself referred in his 2001 article on PB, namely the distinction between genetic enhancement by means of selection and by modification, for example by modifying an already given genetic make-up. The first type of procedure is structurally analogous to choosing a partner with whom one wishes to have offspring. The second type of procedure, however, is structurally analogous to educating one's offspring (Sorgner 2013).

In the following paragraphs I will merely state a few reasons in favour of my claim that selecting an already given genetic make-up and selecting a partner with whom one wishes to have offspring are structurally analogous procedures (Sorgner 2011b, 21–25).

By choosing a partner with whom one wishes to have offspring, one implicitly also determines the genetic make-up of one's child, as 50% of their genes come from one's partner, and the other 50% from oneself. By selecting a fertilized egg, one also determines 100% of the genetic make-up by means of selection.

One objection which might be raised here is that selecting a fertilized egg cell is a conscious procedure, but normally one does not choose a partner according to their genetic make-up such that one has specific genes for one's child. However, a reply to this can maintain that our evolutionary heritage might be more effective during the selection procedure of a partner than we consciously wish to acknowledge. In addition, the qualities according to which we choose a fertilized egg after PGD might not have been chosen as consciously as we wish to believe, but might be influenced more on the basis of our unconscious organic set-up than we wish to acknowledge. It might even be the case that the standards for choosing a partner and for choosing a fertilized egg might both be strongly influenced by our organic make-up and evolutionary heritage, such that both are extremely similar.

A difference between these two selection procedures is surely that in the one case one selects a specific entity, a fertilized egg, but in the other case a partner, and therefore only a certain range of genetic possibilities. However, given the latest epigenetic research, we know that genes can be switched on and off, which makes an enormous difference on the phenomenological level. Hence, it is also the case that by choosing a fertilized egg we choose only a certain range of possible phenotypes of the later adult, as is the case by choosing a partner for procreative purposes.

The aforementioned comparison provides some initial evidence for holding that there is a structural analogy between choosing a partner for procreative purposes and for choosing a fertilized egg cell after PGD. Given PB, and given that structurally analogous procedures ought to be evaluated analogously, PB implies not only that there is a moral duty to select the child with the best chance of the best life, but also that there is a moral duty to select the partner for having offspring such that there is the greatest likelihood that a child with the best chance of the best life can be realized. Anyone who does not stick to this moral duty can be reprimanded for acting immorally.

A moral principle, and PB claims to be such a principle, which implies the moral duty to select a partner with whom one's offspring promises to be best, is violent, and hence immoral.

3.4.3 Conclusion

The main goal of this section was to deal with the question whether we have a 'Moral Obligation to Create Children with the Best Chance of the Best Life?' After having dealt critically with Savulescu's PB, which claims that there is such an obligation, I conclude that his arguments in favour of such a duty fail, as they are inconsistent and immoral. From this it does not yet follow that there is no such duty, but it merely means that his arguments in favour of such a duty are implausible. Without having the space to move beyond this conclusion, I wish to point out that I regard the principle of PA as an appropriate one. Savulescu has argued in various articles that PA is not the appropriate attitude with respect to the process of selection after IVF and PGD. According to Savulescu, PA can be summarized thus: 'Procreative autonomy. If reproducers have decided to have a child, and selection is possible, then any procreative option selected by reproducers is morally permissible as long as it is chosen autonomously' (Savulescu/Kahane 2009, 279).

Given that there is a structural analogy between choosing a partner for procreative purposes and choosing a fertilized egg cell after PGD, I regard PA as morally appropriate for a liberal state. Here, the individual's right to live a good life according to his own concept of a good life is of central importance. During the previous 500 years, enormous and intensive struggles in various social and political fields have taken place, until negative freedom has been widely recognized as a central norm, and I regard it as important to always take this achievement into consideration. A move away from PA and towards PB is a move in the wrong direction, because it introduces new paternalistic structures. This time the structures are not given on a legal level, but merely on a moral one. Still, it has the effect that such structures violently intrude in the private realm of individuals and violently attack the precious achievement that it is widely recognized that a radical multiplicity of concepts of the good can be appropriate. Hence, I finally conclude that not only do we not have a 'Moral Obligation to Create Children with the Best Chance of the Best Life', but I am even bound to claim that it is immoral to defend a 'Moral Obligation to Create Children with the Best Chance of the Best Life', or, in other words, PB is not a moral principle, but an immoral one.

4

A Fictive Ethics

In this chapter I analyse the most important ethical issues concerning transhumanism. Michael Sandel argues that gene modification and gene selections need to be rejected. However, his main reason is not that they are morally wrong, but that they imply vicious character traits. Parents who employ these technologies do not possess the central parental virtue, namely that of unconditional love. I will argue in the first section of this chapter why neither do I share Sandel's communitarianism nor is it the case that his conclusion is a necessary one, given his own premises. Instead, it is more plausible to hold that using gene technologies can demonstrate a parental virtue.

In all of the aforementioned arguments, the question of the good life is a central one. Is any general judgement concerning the good life possible in a naturalist world, a world without a personal God? Is anything permissible, is nothing forbidden, or can anything be said concerning the good life, given these circumstances? There are various transhumanist takes on this issue. In the second section I will show that all the widely used ways of demonstrating transhumanism in the public media are implausible. Superman on Viagra, or Wonderwoman with Botox, is not what all transhumanists subscribe to as a central goal, and not what they should subscribe to either, given their own initial premises. I show why a radically pluralist concept of the good is more plausible. No non-formal judgement concerning the good is plausible.

Section three will be dedicated to the question of what counts as morally right from a transhumanist perspective. Even though any concept of the right is regarded as fictive, this does not imply that it is arbitrary. We do have criteria for evaluating moralities. These criteria are historically and culturally embedded, but this does not mean that they are meaningless. They are meaningful for our lives. Here, I will present central aspects of what a non-anthropocentric, a non-essentialist, and a non-dualistic concept of personhood would have to consider. Thereby, I also distance myself from Singer's suggestion and present a more inclusive alternative.

In section four I further clarify many widely shared misunderstandings concerning transhumanism, as it needs to be noted that there is an enormous diversity of different approaches within this cultural movement. I particularly focus on the issue of whether an affirmation of a utopia is necessary for transhumanists, which I deny. I affirm a non-utopian take on transhumanism. There might not be an ideal future social arrangement, but a consideration of a plurality of concepts of the good and the relevancy of expanding the health span are central guidelines which ought to be taken into consideration in all circumstances.

In section five I am concerned with some wider-ranging issues of transhumanism, and the question concerning the meaning of life. Can there be a transhumanist meaning of life? Is the question concerning the meaning of life meaningful? How does the question on the meaning of life relate to other transhumanist endeavours? I will debunk many widely shared prejudices concerning transhumanism in this section.

The great variety of aforementioned reflections show that my take on transhumanism is a philosophically well justified one, which does not apply to the majority of transhumanist reflections. It is non-utilitarian, non-utopian, and non-linear. It does not imply strong truth-claims. It is also non-anthropocentric, non-essentialist, and non-dualistic like most critical posthumanist philosophies. Hence, it lies in between post- and transhumanism. There is the need to think in between post- and transhumanism, if one doubts the existence of a categorically dualistic ontological anthropology. We ought to think in between post- and transhumanism, and affirm a type of metahumanism instead.

4.1 Using gene technologies as a vice?[1]

The main goal of the chapter is to argue against the plausibility of Michael Sandel's communitarian views on genetic enhancement by drawing upon selected aspects of Nietzsche's philosophy. Thereby, the contemporary relevance of Nietzsche's thinking on community, virtue ethics, and enhancement technologies becomes clear.

This section is divided up into the following three parts. Firstly, I present central aspects of Nietzsche's communitarianism which are particularly relevant concerning the future development of human beings. However, due to his rejection of the norms of freedom and equality, I do not regard his social suggestions as an appealing cultural vision (Sorgner 2010a, 232–239). Secondly, I am concerned with Sandel's communitarianism by dealing with his virtue-ethical arguments against genetic enhancement. Thirdly, I draw upon some of Nietzsche's central insights which I regard as convincing to argue against the plausibility of Sandel's position. Thereby, the relevance of Nietzsche's philosophy for today's virtue-ethical discourses concerning communitarian views on genetic enhancement becomes clear. By drawing upon selected plausible insights of Nietzsche's philosophy, it is possible to develop arguments against Sandel's communitarian views on genetic enhancement which are important for today's discourses in this field.

4.1.1 Nietzsche's communitarianism, virtue ethics, and the overhuman

Nietzsche was affirmative of communities and can be classified as a communitarian.[2] I find Young's reading of Nietzsche in this respect convincing (Young 2006, 1–2). In addition, I regard it as highly important to stress the communitarian aspect of Nietzsche's thought, as it is central for grasping the cultural embeddedness of his philosophy and it is this element of his thinking which had a massive influence upon further German thinkers. Ferdinand Tönnies, who is one of the founders of sociology in Germany, represents a paradigm case and Nietzsche's inspiring reflections concerning social structures are undeservedly neglected in the English-speaking world.

Besides many other books, Tönnies' wrote the monograph *Gemeinschaft und Gesellschaft* (Tönnies 1979), in which he distinguishes a community from a society. A community consists of organic structures which are being united in a shared account of the good and in which the family plays a central role. A society is based upon human beings as atomic individuals whose main interest lies in the increase of their own personal wealth. These overtly schematic and reductive descriptions hint at the possibility of there being an analogy between the community and society distinction and the communitarianism and liberalism division. Tönnies favours communities over societies, like most German thinkers of this time.

Nietzsche's ideal community is a special one, because it consists of two classes whereby the distinction depends on someone's biology, or, rather, psychophysiology, and not on the person's social background. In Nietzsche's favoured type of community there are a few, active artistic and philosophical creators and many, passive workers who provide them with their daily needs (KSA 1, 767). This understanding of Nietzsche's political vision gets further support from his Sipo Matador section, which hints at his favouring a healthy aristocracy as the best way of social organization (KSA 5, 206f). He regards masterly virtues as appropriate for the few leading community members, and slavish virtues for the many supporting members of the community, which are the basic ethical constituents of his political and cultural ideal. His reason for regarding this type of community as superior to other types of social organization is that it increases the likelihood of the coming about of the overhuman, the next step in human evolution (KSA 13, 316f). This insight reveals that Nietzsche's version of communitarianism and his eugenic reflections are closely connected.[3] It has to be noted that here I use the word 'eugenics' in an ahistorical sense.

Concerning Nietzsche's understanding of evolution, I refer to some passages of his notebooks in which Nietzsche develops a basis of a theory of evolution (KSA 13, 316f). There, he explains that by promoting the tension between strong and weak human beings the likelihood of an evolutionary step towards another species increases. At the same time, it has to be acknowledged that Nietzsche was critical of Darwin. Many scholars identify this critical attitude as rejection of an evolutionary approach. Given that Nietzsche attempted to develop guidelines of an evolutionary theory goes against this understanding of Nietzsche's approach. By realizing that Nietzsche also stressed that he attacks those most vehemently who are closest to his way of thinking, for example Socrates (KSA 8, 97), it becomes clear that his attack on Darwin does not have to imply a rejection of an evolutionary understanding of human development. Hence, I do not think that his critical comments concerning Darwin go against my reading of the overhuman as a member of a new species. However, it is the case that Nietzsche criticized Darwin for seeing the will to survive as the basic human motivation, which is based in his regarding the will to power as the underlying motive of human acts (Sorgner 2007, 62).

By considering specific aspects of Nietzsche's philosophy which are not necessary for his communitarian politics, it is possible to argue against Sandel's contemporary communitarian arguments against certain methods of procreation (part three of this section), for example genetic enhancement. This methodology reveals the contemporary relevance of Nietzsche's way of thinking. Thereby, I focus on specific points of Nietzsche's view, like his will-to-power ontology and his theory of aristocratic virtues, and employ and develop them further so that they can be employed in contemporary debates. If I considered Nietzsche's views in general, then a different picture would have to be presented which is closer to a Gattaca-like social structure and hence not a very appealing vision (Sorgner 2010a, 239–242). It is definitely not a vision of which I am in favour, and I am glad that this judgement is widely shared today.

Sandel's communitarianism is founded on his virtue of unconditional love. By drawing upon Nietzsche's will-to-power anthropology and his virtue theory, in particular his masterly virtue of truthfulness, it is possible to argue against Sandel's reflections. My own position, which I develop further by drawing upon selected insight from Nietzsche's philosophy, is anti-communitarian. By developing further Nietzsche's noble virtue ethics, Nietzsche can be used to support an anti-communitarian way of thinking.

I am progressing in this way, as I think that it renders his way of thinking more relevant for contemporary discussions (Sorgner 2010a, 218–237). Some readers might find it odd that I am stressing that Nietzsche affirms a communitarian political philosophy, but I am at the same time drawing upon elements of Nietzsche's thought to reject a communitarian position. However, I regard this methodology as useful for developing timely philosophical insights. By claiming A, I am being true to Nietzsche's point of view. However, by arguing in favour of B, I am making some of his insights important for contemporary discussions. It is one way of showing the contemporary relevance of his way of thinking. Is this an odd methodology? Postmodernists focus on a specific reading of Nietzsche's perspectivism and develop it further. Transhumanists are concerned with one interpretation of Nietzsche's overhuman which they then integrate into their point of view. A partial reception of a philosopher is still one way of thinking within the tradition of a given philosopher.

By stressing the aforementioned premises, it becomes clear that Nietzsche's thinking on community was closely connected to his procreative reflections concerning the bringing about of the overhuman. Yet this does not imply that Nietzsche cares solely about the exceptional types and not about the herd types. A Nietzschean communitarian social structure is in the interest of both the few creators and the many workers, because therein members of both groups can do what corresponds to their psychophysiological constitution. In addition, workers are better off in such a community than in a society, because here the cultural creators care for their supporting workers and thus they have more security than they have in a liberal democracy in which no one cares for the workers once they do not work or are unemployed, according to Nietzsche (KSA 2, 296). This insight represents one central argument in favour of his defending a type of communitarianism.

Concerning Nietzsche's goal of bringing about the overhuman by means of human evolution, Nietzsche explains that by promoting the tension between strong and weak human beings the evolutionary step towards another species can occur (KSA 1, 767). Let me clarify this point further: Nietzsche is in favour of a community of human beings. This community consists of the few cultural creators and the many workers who supply the creators with their daily needs. It is a community because the structure is such that it is both in the interest of the few and in that of the many. Nietzsche is in favour of such a community because it increases the likelihood of the coming about of the overhuman. In that case, Nietzsche is a communitarian with respect to social organizations of human beings. However, this type of community is solely of instrumental value, because it increases the likelihood of the generation of the overhuman.

How can the overhuman come about? In this context the tension between the few and the many becomes relevant. However, it is not the social tension between these two types of human beings which is at work, as then it would not be appropriate to refer to this type of social organization as a community. The tension Nietzsche has in mind is a biological tension, the tension within the boundaries of the human species. According to selected passages of his notebooks, evolution works as follows (KSA 13, 316f). Each species has certain boundaries beyond which the members of the species cannot develop. If the members of a species reach the outer limits of the borders of the species, then the likelihood of the evolutionary step towards the overhuman species is highest. If the few reached the upper limit of the boundaries of the human species and the many were at the lower end of the boundaries of the human species, then there would be a biological tension between these groups of human beings. It is this type of tension Nietzsche is talking about and what I refer to when I use the word 'tension'. I do not mean social or political tensions between the few and the many. Even if such a biological tension exists, the few and the many can live together in one community, which is in both of their interests.

This reading of Nietzsche's social and political philosophy is a communitarian reading, as it implies that the best type of social organization for human beings is the community. However, the community of the few and the many is not the ultimate goal of Nietzsche's conception. The community is of instrumental value, as it increases the likelihood of the coming about of the overhuman. Ultimately, the overhuman counts, as the overhuman is the meaning of the Earth (KSA 4, 14). However, concerning the best type of human social organization, the community of the few and the many is what is best.

According to Nietzsche, a community is best for the few and the many to lead a good life, and to fulfil the meaning of the Earth, that is, to increase the likelihood of the coming about of the overhuman, because a community might be particularly supportive for many human beings to reach the boundaries of the human species such that the conditions are given for the evolutionary step towards the posthuman to occur. In contrast to this type of community, a society implies nihilism, hedonism, and liberalism, which may make it difficult for many individuals to possess qualities or to acquire capacities such that these correspond to the qualities of the boundaries of the human species. Consequently, Nietzsche is an anti-liberal thinker (KSA 6, 106). To increase the likelihood of an evolutionary step occurring, according to Nietzsche, a certain amount of human beings must reach the boundaries of the human species.

Nietzsche's reflections on the development of human beings have been received intensely in various eugenics debates (Stone 2002). This applies both to the history of eugenics as well as to the latest reflections on this topic; for example, Nietzsche's position concerning the 'human zoo', genetic enhancement, and the future of human evolution is discussed intensely by transhumanists, bioethicists, and Nietzsche scholars.[4] The infamous Sloterdjk–Habermas debate is also concerned with this area of topics. In his essay concerning liberal eugenics, Habermas identifies Nietzsche's ethical ideals both with transhumanist as well as with fascist ideas:

> A handful of freaked-out intellectuals is busy reading the tea leaves of a naturalistic version of posthumanism, only to give, at what they suppose to be a time-wall, one more spin – 'hypermodernity' against 'hypermorality' – to the all-too-familiar motives of a very German ideology. Fortunately, the elitist dismissals of 'the illusion of egalitarianism' and the discourse of justice still lack the power for large-scale infection. Self-styled Nietzscheans, indulging in fantasies of the 'battle between large-scale and small-scale man-breeders' as 'the fundamental conflict of all future,' and encouraging the 'main cultural factions' to 'exercise the power of selection which they have actually gained,' have, so far, succeeded only in staging a media spectacle. As an alternative, I will appeal to the more sober premises of the constitutional state in a pluralistic society, as a way of contributing to some clarification of our confused moral sentiments. (Habermas 2003, 22)

Here, Habermas is concerned with Sloterdijk's speech *Rules for the Human-Zoo* (Sloterdijk 2009, 12–28), whereby he does not mention the name 'Sloterdijk' but clearly refers to his speech by literally citing selected passages from it. This citation shows that Nietzsche is still seen as a proponent of dangerous fascist breeding fantasies, because Habermas identifies Nietzschean breeding fantasies with the continuation of a 'very German ideology'. I do think that Habermas is right in stressing that there are certain structural analogies between Nietzsche's philosophy and selected reflections by transhumanists. However, in contrast to Habermas I do not see this as a reason for rejecting all variants of transhumanism or all aspects of Nietzsche's philosophy.[5] However, there are aspects of Nietzsche's philosophy, like his cultural vision (a two-class community), which I find morally problematic and I am glad that I am sharing this evaluation with the majority of members in enlightened countries. Many other aspects of his insights are highly plausible and they can be referred to and employed in contemporary debates without the need to embrace all of Nietzsche's philosophy. In part three of this section, I will consider Nietzsche's will-to-power anthropology (Sorgner 2007, 39–65), and his theory of masterly virtues in particular (Sorgner 2010a, 143–50), as

I regard both theories as convincing and relevant. Some further clarifications concerning these issues need to be introduced at this stage.

As Nietzsche regards the will to power as the fundamental drive underlying all human actions, moral systems and virtues are also a result of this underlying drive. Yet, due to there being a plurality of human types, the will to power expresses itself in a manifold way concerning moralities and virtues, too. Nietzsche's fundamental distinction concerning psychophysiological human types is that between the master and the slave, and one central characteristic of this distinction is that between an active and a passive being.

I am not dealing with the priest-type here (Sorgner 2010a, 135–40), because I am merely summarizing some specific issues which are immediately relevant for this topic. In addition, it needs to be stressed that the distinction between the master and the slave sounds like a simple-minded dualist categorization of the world which cannot correspond to the lifeworld. However, Nietzsche is aware that that is the case. Hence, he points out that there are slavish and masterly traces in all of our characters (KSA 5, 208). He merely uses this categorization in order hint at the differences between human beings in whose character traits one of the two attitudes comes up more often. The following characterization both of the master and of the slave falls short of the complexity of Nietzsche's description. In particular the relevance of the cultural circumstances for the various developments cannot be considered here. However, the following hints reveal what is important in the context of this argument.

The active master lives in accord with his own physiological demands and attempts to realize them so that the will to power can unfold itself in a direct way. The will to power within the reactive slaves, on the other hand, unfolds itself indirectly, which becomes clear if one considers the character traits Nietzsche ascribes to the slave type, for example descriptive and normative equality, a longing for the afterlife with which comes a devaluation of the sensual world, and the claim of the universal validity of one's own values (Sorgner 2010a, 128–134). The master type can be described with the qualities self-affirmation, nobility, and, in particular, truthfulness (Sorgner 2010a, 141–143). By being true to oneself, one's own drives, and also one's own immediate expression of the will to power, the master realizes what is good for him. What is good for him (or her), the master type refers to as good. He is less concerned with all the other activities to which he refers to as bad. However, in both cases he does not claim that what is good and bad for him applies to anyone else besides himself.

The slave type, on the other hand, is reactive, weak, and not inclined towards an immediate realization of his drives (Sorgner 2010a, 193–208). He just realizes that the masters oppress him (or her) and use him for their own ends. As these acts do not allow that his own will to power can express itself immediately, he refers to these acts as evil. However, he does not claim that these acts are evil solely from his own perspective, which is how masters formulate their value judgements; slaves claim that the related acts are universally evil and that such value judgements are supposed to apply to all human beings at all times, and those who do not stick to them will be punished in their afterlife. The main focus of slave morality lies on evil. All the other acts which cannot be classified as evil can be referred to as good.

As both utilitarianism as well as deontological ethical theories are focused on what is universally right and wrong, they can be classified as slave moralities, from Nietzsche's perspective. Consequently, he rejects them (KSA 5, 160). One could object to my referring to Nietzsche's ethics as virtue ethics, because it is also clear that Nietzsche is fairly critical of virtues. However, his criticism applies only to slavish virtues which claim universal validity. He is full of praise for *virtù*, a this-worldly Renaissance understanding of virtues (KSA 12, 518). It would be false to refer to *virtù* as the Renaissance understanding of virtue, because in particular a Neoplatonic virtue ethics whose primary goal is seen in the *unio mystica* and thereby in the return to the one is another widespread Renaissance type of philosophy, which is represented best by Marsilio Ficino and Pico della Mirandola. Nietzsche does not affirm their understanding of virtues. When he talks about Renaissance virtue, he means a specific humanist tradition which concentrates more on the human body, its function, and beauty, and it is this tradition which moves towards a more immanent understanding of human beings, which is closer to his understanding of *virtù*.

It is his immanent understanding of the world which leads Nietzsche to his task of naturalizing ethics. This also applies to his theory of virtues, and he is ready to defend his naturalized understanding of virtue against the preachers of virtues (KSA 2, 517). One's own virtue can be identified with the health of one's soul (KSA 3, 477), whereby the soul needs to be understood as a part of one's body (KSA 4, 39). When he talks about masterly virtues, nobility and truthfulness are mentioned most often by Nietzsche (KSA 13, 474f). Other virtues are enumerated, too, such as bravery or solitude (KSA 12, 506). There are three further features concerning his virtue ethics which are particularly noteworthy. According to Nietzsche, it is better to have one virtue rather than two (KSA 4, 17), it is virtuous to also send one's virtues to sleep at the right time (KSA 4, 33), and what it means to be noble is bound to take different forms in men and in women (KSA 6, 63). It also seems to me that there is a close connection between truthfulness and nobility in Nietzsche's philosophy, and whoever listens to his own drives, and lives accordingly, embodies *virtù*, because in them the will to power can find its immediate expression.

The slavish virtues, on the other hand, are supposed to be universally valid and represent qualities such that they are in the interest of all those people who do not possess them, according to Nietzsche, and weaken those who have them (KSA 3, 391). Sympathy is one of the most famous virtues in question. Still, the same applies to altruism, faith, love, and hope, or humbleness and chastity. As these virtues weaken the virtuous, Nietzsche regards them as slavish virtues (Sorgner 2010a, 145–146). These virtues are in the interest of the slaves because the virtues weaken the masters, and by possessing the slavish virtues the masters diminish their own power. Again, there is a close connection between the virtues and Nietzsche's will-to-power ontology.

By relating virtues to his will-to-power ontology, Nietzsche fulfils his task of naturalizing morality, because his will-to-power ontology is a kind of naturalist understanding of the world. Within the two-class community for which Nietzsche argues, the few active artists and philosophers can live in accordance with their masterly virtues, and the many workers can hold on to their slavish virtues. Both can live in accordance with their own psychophysiological demands and therefore live good lives without there necessarily being a social tension between the two classes involved. Given this way of social structure, the biological tension within the human species becomes intensified. Thereby, the likelihood might be increased of promoting the development towards the overhuman who belongs to a new species and represents the meaning of the Earth. Hence, Nietzsche's two-class community promotes the good life of the active and passive members of the community and provides them with a meaning for their lives by having a goal which transcends their own existence, that is, the process of the coming about of the overhuman.

A healthy aristocracy is in the interest of the few masters, because being in charge and being active corresponds to their psychophysiological structure, and it is also in the interest of the many slaves, because being led but also being embedded in a safe social organization is what corresponds to their passive psychophysiological foundations. Nietzsche's anthropology is descriptive in so far that it implies that there are active and passive human beings. However, it is also normative in the sense that it ought to be the case that the active ones are the governing ones and free creators and the passive ones are there to provide the others with the means for living a life dedicated to artistic creativity. The concept 'normative' here must not be read in the sense that Nietzsche makes a judgement concerning what is universally right. This is not the case. It is a normative position in the sense that he makes a statement concerning the good. All masters (slaves) are active (passive) and ought to live in accordance to their being active (passive) to live a good live. Hence, living actively (passively) promotes their living a good life. The connection between their being active or passive and the normative judgement that they ought to live in accordance with their personality is not a logically necessary connection but, rather, a prudential one. The slavish personality type is passive and, hence, it is prudential for them to consider their own passive personality type, because in this way they increase their likelihood of living a good life, as whoever acts in accordance with their own drives lives a good life. By focusing on the concept of the good life, Nietzsche prioritizes the good over the right, which is characteristic of communitarian philosophies. Still, it is not communitarian in a traditional sense in so far as he does not give a single account of the good for which he claims universal validity. What he does put forward is a formal account of the good which implies that it is in the interest of or valuable for members of each type of personality to live a life which corresponds to their own psychophysiological demands. Hence, his ethics encloses the central traits of a communitarian position, because it prioritizes the good over the right and aims for a social structure which is in the interest of all its members, whereby he stresses that the interests of human beings differ depending on their having a master or slave type of personality. In addition to this aspect of his position, Nietzsche's central reason for upholding it is that it helps human beings to move beyond themselves, so that the likelihood of the coming about of the overhuman increases, which also provides them with meaning in their lives (Sorgner 2010a, 209–210).

4.1.2 Sandel's virtue-ethical rejection of genetic enhancement

Concerning contemporary debates, I focus on Sandel's communitarian suggestions regarding the question of genetic enhancement, as, like Nietzsche, he is a communitarian virtue ethicist who is concerned with questions of procreation. In this section I critically describe Sandel's communitarian views on genetic enhancement. Sandel stresses explicitly that he doubts that currently dominant moral standards provide a reliable basis for rejecting methods of genetic enhancement: 'An ethics of autonomy and equality cannot explain what is wrong with eugenics' (Sandel 2007, 81). However, by referring to a virtue-ethical approach he believes he is able to explain what is problematic concerning genetic enhancement. Thereby, he is primarily concerned with two heteronomous versions of genetic enhancement, namely genetic enhancement by selection and genetic enhancement by modification. However, he also attempts to provide reasons against the option of autonomous genetic enhancement, that is, the case where a person decides for himself to be genetically enhanced: 'If bioengineering made the myth of the "self-made man" come true, it would be difficult to view our talents as gifts for which we are indebted rather than achievements for which we are responsible' (Sandel 2007, 86–87). However, this line of thought is not plausible; as such an either/or situation implies a categorical duality between the option that we are either solely indebted or solely responsible. Most probably, there has never been a historical situation in which human beings have been solely indebted concerning their talents. Without education or any kind of transformative process, which is usually dependent on some kind of technology, no human being would possess any significant talent. Talents need to be developed and developmental processes need engagement and, hence, we have always at least partly been responsible for the development of our own talents. Given the use of autonomous genetic enhancement, this situation would not change in principle. We would have some talents which we have received via our initial genetic make-up and we have the option of developing them further – either by means of self-transformative technologies or by means of genetic-alteration technologies. In both cases, it is the case that technologies are involved. Furthermore, using a genetic enhancement technology is not categorically different to the use of a traditional technology. Consequently, I regard this argument of Sandel's against the use of autonomous genetic enhancement as weak.

In any case, the main focus of his reflections was heteronomous variants of genetic enhancement. In this context, he addressed the relevance of the question of the *telos*, the goal, of the activity in question: 'Arguments about the ethics of enhancement are always, at least in part, arguments about the telos or point, of the sport in question, and the virtues relevant to the game' (Sandel 2007, 38). He also addresses the issue of genetic enhancement in the field of sports by considering the *telos* of the activity in question (Sandel 2007, 29). The *telos* of sports is another complex issue, which is the reason why I will immediately focus on the central issue at stake, namely that of parenting.

Concerning parenting, Sandel puts forward several narratives which are supposed to support the position that parents ought to accept their children as they come, in contrast to the practice of choosing our friends and spouses, whom we choose at least partly on 'the basis of qualities we find attractive' (Sandel 2007, 45). This acceptance is the *telos* of parenting, according to him, and it implies the basic stable attitude of the 'openness to the unbidden' (Sandel 2007, 45) which he sees as central concerning the parents' relationship towards their children. He does not explicitly specify a reason for this being the case. However, given the context of this judgement, it seems plausible to read Sandel such that it is founded on the following insight: 'Parent's love is not contingent on the talents and attributes the child happens to have' (Sandel 2007, 45). It has to be noted that he writes 'parent's love is', as if this sentence was a descriptive one. However, that this cannot be the case becomes clear, given the enormous amount of cases where parents abuse, torture, and kill their own children. His phrase conceals a moral statement which Sandel wishes to defend: parental love ought to be independent of the talents and attributes the child happens to have. Hence, virtuous parents do not use genetic enhancement technologies to select or modify the talents or attributes of their children.

This general position seems highly problematic, because it implicitly stresses that there is a necessary conflict between the virtue of unconditional love and the use of genetic enhancement technologies. Sandel's definition of unconditional love implies accepting the talents and attributes of the other as they are. However, to simply accept the other as who he or she is implies that education would not be an appropriate action either. Yet this is not the case, and this is not what Sandel has in mind either. Hence, further qualifications need to be considered.[6]

The openness to the unbidden implies that the option of choosing a fertilized egg after IVF and PGD is not one that a virtuous parent would choose. The procedure in question can be described as genetic enhancement by selection. Furthermore, it seems to be the case that education or genetic enhancement by modification would not be options either, because thereby parents promote, develop, and enhance the talents and attributes of their children. Let me analyse each of these two cases and start with the latter one. According to Sandel, the aforementioned inference is not one he affirms wholeheartedly. When dealing with education, he introduces the distinction between accepting and transformative love as two manifestations of unconditional love (Sandel 2007, 49–50). It can be the case that parents' transformative love demands to shape and direct the development of their children, as this behaviour can be virtuous. Again, Sandel's main focus is the virtue of unconditional love. He argues that transformative love is one way in which unconditional love can manifest itself.

To uphold the validity of this argument, it seems to imply that Sandel has to uphold that genetic enhancement by modification and traditional educations are not structurally analogous procedures, because otherwise his argument would seem to be inconsistent, because then he would affirm that certain types of shaping and directing a child's life can be virtuous, but also that all types of genetic enhancement by modification are vicious. Yet, if Sandel upholds that the two procedures in question are not parallel ones, then his argument rests on a highly implausible premise, as there are several reasons for claiming that traditional education and genetic enhancement by modification are structurally analogous procedures.[7] Still, Sandel is aware of this option and accepts that traditional education and genetic enhancement by modification can be seen as structurally analogous procedures. Nevertheless, he rejects the argument that this is a reason for accepting genetic modifications, because he claims that genetic enhancement by modification has to be analysed as an expression of hyperparenting, which has to be in conflict with the norm of unconditional love (Sandel 2007, 52) and represents a problematic attitude which is characteristic of our times (Sandel 2007, 62). Here a brief reply has to be mentioned: it is highly dubitable that all types of genetic enhancement by modification have to be identified with hyperparenting. There are plenty of reasons for claiming that genetic modification and traditional education are parallel procedures (see Sorgner 2010a, section 1). Furthermore, it can be the case that both procedures (education and genetic modification) aim for the same goal, for example the increase of mathematical talent, which, for example, could be realized by means of either early childhood education or genetic modifications. If this talent is promoted by education, this is seen as a task which is being fulfilled by virtuous parents. Why should parents turn vicious, if the same goal is being promoted by a different means, for example genetic modifications? I take it that genetic modification technologies are merely a means for promoting specific goals, for example cognitive capacities. The same goals can also be promoted by educational measures, in which cases they are mostly seen as appropriate ones. What is so bad about using genetic modification technologies that the parents who do so are regarded as vicious? Sandel does not have a plausible reply to this question. He could claim that gene technologies are not solely a means but alter the relationship between humans and their environment. Again, it could be replied that the same can be said of educational measures, as educational measures have always brought about genetic modifications, if the latest research in the field of epigenetics is taken seriously. Hence, his line of argument on this issue is complex and interesting but is not plausible.

Let us come back to the first case in question, the case of genetic enhancement by selection. An immediate challenge concerning Sandel's position becomes clear, given his position of seeing health as an intrinsically valuable quality (Sandel 2007, 48). Thereby he weakens his own line of thought. If it is the case that health is intrinsically valuable, then there is a clear reason for choosing a child after IVF and PGD, because thereby the likelihood of the child's being healthy can be increased.

A further argument Sandel advocates stresses that caring teaches parents the openness to the unbidden, which is worth affirming, according to him, as it enables us to live with dissonance, which otherwise would not be possible for us (Sandel 2007, 86).

This argument is worse than the earlier one, as it includes a non sequitur. The attempt to make the best of a situation or a character trait is logically independent of one's attitude concerning dissonance. I can try to create a musical piece like Nyman and realize that it will end up being rather like a Henze piece. However, this does not have to mean that it will be impossible for me to live with it. This analogy does not hold, but it helps me to hint at the understanding that Sandel's suggestion is far from plausible.

In addition, it needs to be stressed that his line of thought, which leads to the rejection of genetic enhancement by selection, is not plausible, because there is a structural analogy between selecting one's partner for procreative motives and selecting a child after IVF and PGD.[8]

As it is the case that by choosing a partner with whom one wishes to have offspring, one thereby implicitly also determines the genetic make-up of one's offspring it is not a categorically different act to choosing a fertilized egg after IVF and PGD. However, Sandel makes this judgement which seems highly implausible. Furthermore, he stresses that the *telos* of parenthood is connected to the 'openness to the unbidden'. I wonder why there is supposed to be any value in the 'unbidden'. It seems as if the unconditional love of parents which Sandel mentions in the same paragraph implies the 'openness to the unbidden'.

In addition, in the aforementioned arguments Sandel refers to the domain of social challenges related to genetic enhancement (Sandel 2007, 92), wherein he stresses the following two arguments in particular: (1) if we give up the notion of giftedness, we will be on the way towards a totalitarian social structure; (2) if we give up the notion of giftedness, the strong will reject the norm of equality. Yet, neither of these dangers is necessary (see 'Zarathustra 2.0 and Beyond', Sorgner 2011b). The social structures depend upon what we fight for, and what we manage to realize as such structures. The notion of the strong is problematic in principle, as we all are weak in some respect and at certain times, for example when we are old, ill, or simply asleep.

So far, I have merely provided some specific reasons against Sandel's line of thought. In the following third section, I focus on the virtue-ethical communitarian basis of his argument, which rests mainly on the claim that unconditional love ought to be the dominant virtue concerning parenting.

4.1.3 A Nietzschean critique of Sandel's criticism of genetic enhancement

In this section I draw upon selected central insights of Nietzsche's thinking to argue against Sandel's communitarian views on genetic enhancement. Thus, it becomes clear that even though Nietzsche argues in favour of a communitarian ideal which is founded in his views concerning human evolution towards the overhuman, it is possible to draw upon selected insights of his philosophy to argue against contemporary communitarian views on genetic enhancement, in this case the views defended by Michael Sandel.

Sandel's central virtue concerning parenting is that of 'unconditional love'. Thereby, he belongs to a long tradition of philosophers who uphold love as central a virtue which goes back to St Augustine and St Paul. Augustine stresses to love and do what you will in his seventh homily on the First Epistle of John (paragraph 8). Paul in his letter to the Corinthians also points out the central importance of love: 'But now abide faith, hope, love, these three; but the greatest of these is love' (1 Corinthians 13: 13). However, you do not have to be a Christian in order to uphold the central importance of the virtue of 'love'. Feuerbach and, by developing further Feuerbach's position, Richard Wagner also agree upon the immense relevance of love. The complex interplay between power and love in Wagner's *Rheingold* is particularly fascinating (Sorgner 2008) and it is this aspect which makes Wagner's works interesting and relevant for our times (Sorgner 2013c), even though it has to be interpreted in a weakened form in order to make his positions plausible, given today's reception (Sorgner 2014a). Wagner's theory of the music drama, on the other hand, is too romantic and overtly integrates the perfect music drama in an idealized state in which a community is present which is being unified by agreeing on a stronger ideal of the good. I regard a community as a highly problematic ideal (Sorgner 2011a). This doubt of mine concerns not only Wagner's understanding of a community, but applies also to Nietzsche's and Sandel's communitarian visions, too. However, regarding the interplay between love and power, which Wagner presents, the issue is different, and I regard it to be an important contribution for contemporary discussions, too. Sandel and Nietzsche focus on one aspect solely: Sandel on love and Nietzsche on power. By specifically analysing Sandel's argument against genetic enhancement by selection in the following paragraphs, I point out why it is problematic and why it is important to also take Nietzsche's thoughts concerning virtue theory into consideration.

Against the use of genetic enhancement by selection Sandel argues as follows: the central virtue concerning parenting is unconditional love. The parental virtue of unconditional love leads to the appreciation of children as gifts, which is also supposed to imply 'to accept them as they come', which is contained in the attitude of the 'openness to the unbidden'. Hence, a virtuous parent does not select and reject a fertilized egg after IVF and PGD, because such an action is inconsistent with the 'openness to the unbidden' and hence the parental virtue of unconditional love.

Let us assume that this is a valid argument. Firstly, it can be asked whether unconditional love is actually the sole virtue concerning parenting. Secondly, Sandel's position can be challenged by questioning whether it is consistent to hold that 'unconditional love' is consistent with 'transforming love', and Sandel himself stresses that during the process of education both 'accepting love and transforming love' are important. If transforming love aims at the alteration of a child, then there are reasons for claiming that this type of love is not unconditional, because it aims for an alteration of the child and does not accept it as it is. Thirdly, I already hinted at the possibility that transforming love can find its expression in educative measures, but, given the structural analogy between education and genetic enhancement, there is no reason why transforming love should not be able to manifest itself in genetic enhancement by modification. Given that the structural analogy I hold is correct, then there is no reason to reject that transforming love can manifest itself in genetic enhancement by modification. As education does not have to be morally objectionable, genetic enhancement by modification does not have to be morally objectionable either. It needs to be kept in mind that it is not necessary that genetic enhancement by modification has to be seen as an expression of hyperparenting.

As I pointed out before, there are several reasons for doubting the plausibility of Sandel's argument. Now I focus on the central question whether unconditional love is actually the sole virtue concerning parenting. Thereby, the consequences of parenting on the basis solely of unconditional love can be considered. I am not claiming that it is impossible to lead a good life if one's parents are solely determined by unconditional love, but I think that there is a high likelihood that the child of parents solely possessing unconditional love will be spoiled, incapable of dealing with many real-life problems, and not responsible for taking care of his own future. The following argument specifies this position further: the narcissist personality disorder argument. I will draw upon 'The Oblomov case' to further explain the challenge in question.

The narcissist personality disorder argument takes psychological research into personality disorders seriously. According to Millon (Millon 2011, 375–422), the narcissist personality disorder can have two reasons: (1) a lack of love during childhood; (2) too much unconditional love during childhood. The second reason implies that parents always praise their children, hold the child's talents in high esteem, and try to fulfil all of the child's wishes. According to Millon, it is likely that children who were brought up in this way have an overly high understanding or picture of their own selves. As a consequence of the immense admiration of his parents, the child fears that he is unable to fulfil the claims or demands of others, because he thinks that everyone has such high expectations of him.

This attitude can lead to various states. Millon (Millon 2011, 375–422) distinguishes six types of narcissism. Often, a narcissist personality disorder leads to depression because of the tension between the subject's high self-picture, the pressure connected to the fear of not being able to fulfil the supposedly high expectations of others and, as a consequence, the unwillingness to participate in any situation in which his capacities can be tested. However, besides depression, auto-aggression, aggression in general, hyperactivity, and suicidal tendencies are often associated as symptoms with narcissist personality disorder. Without being able to spell out in detail the various facets of narcissism from a psychological perspective, this short description is sufficient for having a reason to argue that unconditional love might not be the sole proper virtue for parenting.

It might be interesting to note further that Nietzsche's analysis of human beings being determined by the will to power can also be seen as a symptom of having suffered from a narcissist personality disorder (Sorgner 2010a, 233–238). He continually analyses that there is a gap between his ideal and his real self which he is trying to fill. Hence, the will to power, the will to gain further capacities, and to move beyond his current state is permanently dominant in him. However, as Millon mentioned, too, this type of personality disorder can also express itself in different forms. A literary example which exemplifies a case of too much love can be found in the novel *Oblomov* by Ivan Goncharov. In this case, Oblomov's parents are described as being full of love for their son and dedicating themselves completely to his education. Too much parental love brings about an enormous passivity without Oblomov wishing to take on any responsibility, whereby his main focus is on his daily nap in the afternoon. As he never learned to fulfil any duties, he tries to postpone any work to another day.

Reflections on narcissism are particularly relevant for doubting Sandel's suggestion that unconditional love ought to be the sole virtue concerning parenting. In the case mentioned here, this attitude leads to a severe personality disorder in the child, which made it difficult if not impossible for him to lead a flourishing life.

The virtue of unconditional love fails to respect the fact that we are living in a competitive, naturalist world, in which we need to become strong and fight for ourselves so as to be able to flourish and to find and defend our own position in the world. It is the naturalist, evolutionary world full of power-driven organisms which Nietzsche describes when he analyses the various psychological structures of human beings, and the relevance of the will to power for all of our acts and even our values and virtues.

It is this naturalist and evolutionary worldview which, after Darwin and Nietzsche, has gained wider acceptance in Western societies in the 20th and 21st centuries. Evolutionary thinking has only gradually entered the various fields of academic disciplines, for example evolutionary epistemology, psychology, ethics, and, in particular, aesthetics are fairly new academic disciplines, and they are steadily gaining more and more acceptance. Due to the disrespect for the relevance of naturalist processes which is implicit in the virtue of unconditional love as defended by Sandel, it is advisable to turn to or at least consider and reflect upon Nietzsche's ethics, whose main aim was to naturalize traditional values and virtues and hence to provide a more plausible basis for them. As a consequence of this goal, Nietzsche suggested that the traditional slavish virtues are highly dangerous, because they do not support and promote one's self interest but instead they are in the interest of the people who do not have those virtues or do not stick to them (KSA 3, 391). Unconditional love, for which Sandel argues, is such a slavish virtue, and in the preceding three examples it becomes clear that this virtue is neither necessarily in the interest of the parents nor in that of the child to whom it is directed.

It might be advisable to turn to Nietzsche's virtue of truthfulness, which is of value according to him because it allows and promoted the immediate realization of his actual will to power. By considering and accepting the immediate expression, his will to power does not imply that everyone strives for the same goals and simply desires to be politically dominant (Sorgner 2007, 58–61). As shown in various other places, his will to power implies a perspectival theory of power, whereby power is not solely based on one single criterion but the criterion for power is dependent on the psychophysiology of each individual human being. Hence, power can be connected with a great plurality of criteria, and consequently parenting can be linked to a wide range of goals. However, what is important in this respect is that not solely the virtue of unconditional love is dominant during the process of parenting but also the virtue of truthfulness, and thereby a certain consideration of the importance of power. Parents ought not to solely affirm every action of their child but also to demand something whereby it depends on the parents' truthful analysis of their own self-understanding what they demand. This is what Nietzsche's virtue ethics can teach us concerning parenting, and what also needs to be considered concerning the question of genetic enhancement by selection.

Sandel argues that, according to the central importance of the virtue of unconditional love, parents ought to be open to the unbidden and decide not to select a fertilized egg after IVF and PGD. However, given the aforementioned psychological research, it seems likely to hold that there is a great danger that the sole attitude of unconditional love can lead to severe personality disorders in one's child. Instead, it might be advisable to accept the importance of power and to consider Nietzsche's insight concerning the relationship between the will to power and his virtue ethics, which leads to the advice that we ought to be truthful to ourselves to make the decision whether we wish to select a fertilized egg after IVF and PGD or not. However, we have several reasons contesting Sandel's position, namely that it does not have to be vicious to use the option of selecting a fertilized egg after IVF and PGD. With Nietzsche it seems more plausible to stress the importance of truthfulness concerning this decision.

If there is a virtue which ought to be considered, the virtue should be truthfulness. Yet, this attitude leads to a great diversity of different attitudes, which again undermines the initial starting point of any virtue ethics, as a virtue ethics presupposes shared virtues and vices within a community. If someone does not possess virtuous attitudes, that person is excluded from the community. You need to have basic stable attitudes which are shared, to be part of a community. Such a social structure undermines human flourishing, due to the diversity of psychophysiological demands, which is the main reason why any community has to be paternalistic, and why I regard a version of liberalism as most in tune with the greatest plurality of personal flourishing.

Besides truthfulness, I mentioned mindfulness and impulse control as possible transhumanist virtues. Yet these virtues are not meant to be basic stable attitudes which necessarily help everyone to lead betters lives, which used to be the traditional understanding of a virtue. These basic stable attitudes are suggestions which might be helpful for quite few people. They are suggestions, which are worth testing out. Truthfulness, mindfulness, and impulse control take the radical diversity of concepts of the good seriously and can be applied to a great diversity of different lifestyles.

4.1.4 Conclusion

As Nietzsche's reflections are of central relevance for reflections concerning genetic enhancement and, like Sandel, he argues in favour of a virtue-ethical version of communitarianism which is relevant concerning the question of the future of human evolution, I apply selected aspects of his position to contemporary discourses. It is important for me to solely refer to some of his insights, as I reject the political visions of both Nietzsche and Sandel due to their potentially paternalistic implications, which I have dealt with in detail elsewhere (Sorgner 2010a, 240).

According to Sandel, the parents' use of genetic enhancement by modification or by selection represents their vicious personality, because it shows that their parenting is not solely determined by the virtue of unconditional love. By drawing upon Nietzsche's virtue ethics, which is closely connected to his will-to-power ontology and selected insights from psychological research, I provided reasons for holding that Sandel's suggestion is implausible because there is a strong risk that the virtue of unconditional love towards their child will lead to a severe case of personality disorder. Sandel's attitude does not consider sufficiently that we are part of a naturalistic world in which it is important for the child to also take responsibility for himself, work at his own capacities, and promote his own flourishing. Nietzsche's virtue ethics takes this aspect seriously because it is closely linked with his evolutionary and naturalistic understanding of human beings. Consequently, there some reasons for holding that the virtue of truthfulness which Nietzsche affirmed can also be seen as an important virtue concerning parenting. Given this virtue, it follows that it does not have to be the case that the option of selecting a fertilized egg after IVF and PGD has to be seen as a choice made by a vicious parent. Hence, genetic enhancement does not have to be seen as morally problematic.

Furthermore, it can be asked whether unconditional love or truthfulness ought to be seen as the better parental virtue. From Sandel's and Nietzsche's perspective the replies are clear. However, maybe Richard Wagner's position (Sorgner 2008, 194–214) is even more plausible, because both in his theoretical writings as well as in his music dramas it becomes clear that an interplay which takes into consideration an appropriate balance of both love and power is the most plausible solution. Reflecting upon and being confronted with the *Ring of the Nibelung* might be particularly helpful for this purpose. Attending a performance of this music drama enables you to be emotionally involved in, but also intellectually separated from, the actions, which are helpful tools for developing a more critical stance concerning the appropriate relationship between love and power. *Katharsis* is connected with pity and fear, according to Aristotle. Pity implies a certain distance from what occurs on stage, that is, your being able to identify with the protagonist due to some similarities, but still being distanced from the protagonist. Fear, on the other hand, implies that you see yourself as the protagonist and experience the fear which the protagonist experiences. It is this hybrid stance which enables the interplay between feeling and thinking to realize a better-informed grasp of the issue in question.

4.2 Three transhumanist types of human perfection[9]

Transhumanists affirm technologies in order to increase the likelihood of the coming about of posthumans. A strong version of transhumanism affirms that there is a moral but not a legal obligation to use specific enhancement techniques. The bioliberal Julian Savulescu, who is not a transhumanist but closely associated with this movement, represents this perspective well (Savulescu 2001, 5–6). Another strong version of transhumanism holds that enhancement technologies necessarily promote good lives of all human beings, and it is this insight which ought to be legally relevant. Aubrey de Grey can be seen as representative of this tradition (de Grey 2007, 335–339). A weak version of transhumanism, on the other hand, might affirm that enhancement techniques increase the likelihood of many people leading good lives without this insight implying necessary legal or moral duties, but it does have consequences concerning moral and legal rights (Sorgner 2013).

Besides there being differences concerning strong and weak versions of transhumanism with respect to moral and legal duties, different concepts of the good life also need to be distinguished. In this section I will deal with three concepts of the good life and then argue in favour of one of the concepts. The concepts I will deal with can be described as follows: (1) the Renaissance ideal: it is represented well in an article by Bostrom (2001); (2) a common-sense account of the good: it can be associated with Savulescu (2001; together with Kahane 2009); (3) a radically pluralist concept of the good: in such an account it is possible to find many Nietzschean traces, and it is also the concept which I regard as plausible (2010a). I will outline the various concepts and thereby I will also analyse the plausibility of the various positions.

4.2.1 The Renaissance ideal

The Renaissance ideal is one which one can find in various authors and it is one which often gets associated with transhumanism within public debates. Upon closer consideration, it needs to be noted that it is rather difficult to find thinkers who uphold this position over a long period of time. Bostrom argues for this ideal (Bostrom 2001); however, it needs to be acknowledged that he has altered and weakened his position on this question in his later writings (Bostrom 2009). Still, in his 2001 article 'Transhumanist Values', he clearly writes the following: 'Transhumanism imports from secular humanism the ideal of the fully-developed and well-rounded personality. We can't all be renaissance geniuses, but we can strive to constantly refine ourselves and to broaden our intellectual horizons' (Bostrom 2001). Two things become clear: Bostrom sees transhumanism in the Enlightenment secular humanist tradition, and he claims that it is this tradition from which transhumanism is supposed to have taken over the concept of the 'fully-developed and well-rounded personality' (Bostrom 2001), which then was developed further by integrating it into a transhumanist framework. It implies that all human beings wish to become 'Renaissance geniuses' (Bostrom 2001), even though it is also clear for most humans that this is not a feasible option. Still, by being aware of this wish and acknowledging its existence and relevance, Bostrom holds that we as human beings can nevertheless try to 'strive to constantly refine ourselves and to broaden our intellectual horizons' (Bostrom 2001). Here, the concept of transhumanism being a hyper-humanism comes out most clearly. However, this is not the case with respect to other transhumanist thinkers.

Even though Bostrom upholds such a strong concept of the good, there is no doubt that (most) transhumanists (including him) argue on the basis of a version of liberalism concerning political issues. The varieties of liberal attitudes among transhumanists are very wide, and one can both find libertarian thinkers, like Max More, as well as more social-democratic ones, like James Hughes. It is also the case that there does not have to be a conflict between holding a strong concept of the good life and acknowledging the relevance of political liberalism, which allows people to act according to a great variety of attitudes. The same tension can be found in the position of Savulescu, who also upholds a strong account of the good, which, due to his utilitarianism, implies rather rigid moral duties, but at the same time argues in favour of a rather libertarian political system (Savulescu/Kahane 2009, 5).

Interestingly enough, it is this type of tension which can also be found in Nietzsche. On the one hand, he stresses that the physiologies of all human beings differ radically from each other and that this has an enormous influence on what is important for living a good life for all human beings. On the other hand, he stresses the relevance of the classic ideal which he associates with great strength. The following three passages from his *Nachlass*, his writings not published by himself, provide some hints concerning the relevance and meaning of the classical ideal in Nietzsche's philosophy, even though it must be acknowledged that his understanding of the classical does not refer to the concept of the good life only, but applies to various other realms too, like the realms of cultural, artistic, and literary creations: 'The highest type: the classical ideal – as the expression of the well-constitutedness of all the chief instincts. Therein the highest style: the grand style. Expression of the "will to power" itself. The instinct that is most feared dares to acknowledge itself' (Nietzsche, 1968 [hereafter WP] 341); 'The classical style is essentially a representation of this calm, simplification, abbreviation, concentration – the highest feeling of power is concentrated in the classical type' (WP 799); 'Classical taste: this means will lead to simplification, strengthening, to visible happiness, to the terrible, the courage of psychological nakedness' (WP 868).

Both in the case of Nietzsche as well as in that of some transhumanists, what constitutes the Renaissance genius or the classical ideal is a 'fully-developed and well-rounded personality' which attempts 'to constantly refine' itself (Bostrom 2001); this can further be associated with the following qualities: 'calm, simplification, abbreviation, concentration – the highest feeling of power' (WP 799), an 'expression of the well-constitutedness of all the chief instincts' (WP 341) and a 'will to simplification, strengthening, to visible happiness' (WP 868). Translating these various aspects into a more accessible language, the ideal is characterized by bodily beauty, strength and great health, intellectual and cognitive excellence, enormous philosophical wisdom and knowledge, and musical and artistic sensibilities and capacities, which coexist in a balanced and integrated personality and come along with a beautiful, strong, and healthy body. In other words, the Renaissance genius can be identified with Albert Einstein with the body of Arnold Schwarzenegger, the face of Johnny Depp, and the wisdom of Aristotle. At least the male version of the ideal can be characterized thus. A female

Renaissance genius might be Marie Curie with the body of Heidi Klum, the face of Kate Moss, and the wisdom of Hildegard of Bingen.

The implication of such a concept of the good is that it would promote the likelihood of living a good life for all human beings, if someone continuously tried to enhance her own capacities with respect to the various aspects of their existence. By means of sport, one's bodily beauty and strength could be trained. By means of music making, one's capacity of making and appreciating music would be promoted. By means of memorizing poems, one's memory would be improved. This list could be continued with respect to all the other aspects mentioned before which are associated with the Renaissance genius. However, not only by means of drawing upon traditional technologies with respect to the qualities mentioned earlier would the likelihood of someone's leading a good life increase, but also by drawing upon emerging technologies the likelihood of approaching the Renaissance ideal can be promoted. Morphological enhancement, for example plastic surgery, pharmacological enhancement, for example the use of Ritalin, genetic enhancement, and cyborg enhancement, for example the use of imaginary I-brains or brainstations, could all be appropriate means for constantly refining oneself, so that one comes closer to the ideal of turning into a Renaissance genius.

4.2.2 A common-sense account

A seemingly weaker account than the first one is the common-sense account of the good life presented by Savulescu. However, it is also a strong account of the good for several reasons. Firstly, it is uncontroversial for him that there is such a thing as the best life: 'A common objection to the PB is that there is no such thing as a better or best life. It is hard to defend such a claim' (Savulescu/Kahane 2009, 278). PB is an abbreviation for his principle of procreative beneficence, with which I have dealt earlier. Furthermore, it is clear to him that we regularly rank the goodness of lives on the basis of the right criterion: 'there are plenty of cases where we can rank the goodness of lives. We do so in numerous moral decisions in everyday life' (Savulescu/Kahane 2009, 279). He even stresses that the concept of the good life is based upon an uncontroversial common-sense consensus which we are already supposed to apply in many everyday situations:

> But although there is this philosophical disagreement, there is considerable consensus about the particular traits or states that make life better or worse, a consensus that would rule out many procreative choices as grossly unreasonable [...] PB doesn't rely on some special and controversial conception of well-being. All it asks us is to apply

in our procreative decisions are the same concepts we already employ
in everyday situations. (Savulescu/Kahane 2009, 279)

Given his conviction concerning our being aware of the appropriate criteria
of a common-sense concept of the good life, it is unsurprising that Savulescu
also specifies them further. However, in this context it turns out that his
concept is not as clear and specific as it seems, given his prior utterances.

He stresses that it is bad to have a disposition for depression (Savulescu/
Kahane 2009, 281) and to have a disability (Savulescu/Kahane 2009, 286). It
is good, on the other hand, to have good memory, and a strong intelligence
(Savulescu 2001, 420), according to his 2001 article. In his 2009 article, having
a strong memory and being able to concentrate well and understand other
people's feelings seems central for him, as he holds: 'How can the capacity
to remember things better, concentrate longer, be less depressed, or better
understand other people's feelings have the effect that one will be less likely
to achieve the good life?' (Savulescu/Kahane 2009, 284). Having a high
intelligence is also helpful for a good life, according to him: 'If parents could
increase the prospects of future children's lives by selecting children who are
far more intelligent, emphatic or healthier than existing people, then PB
instructs parents to select such future children' (Savulescu/Kahane 2009, 290).

I am not entirely clear about all of the implications of his theory of the
good. Does he mean that a human entity is always better off with a higher
intelligence, a stronger memory, or more intensive capacity to concentrate?
I am doubtful whether this is actually the case. I am also grateful for having
a good capacity to forget things. If I permanently remembered all the bad
things friends have done to me or were conscious of dangers related to
what I am doing, then I would most probably decide to move away from
civilization and live like a hermit on the top of a mountain. Hence, I would
be hesitant to claim that having a better memory is necessarily always a good
thing and that we have a moral reason for always choosing the fertilized
egg with the better memory after IVF and PGD (this is what Savulescu
argues). Likewise, it can be asked whether it is always in the interest of a
child to have a higher intelligence. If the child is the only one with such a
high intelligence, while all the people around him are fools, then I would
be hesitant to claim that having such a high intelligence is necessarily in the
best interest of the child, as this capacity might have the consequence that
the child will be excluded from social life. Hence, it seems to me that the
social settings are of central relevance for the qualities necessary for leading
a good life.

Savulescu seems to have realized and acknowledged the impact of this line
of thought between 2001 and 2009, especially with respect to the question of
disability, which can and ought to be transferred also to other domains of a

good life. In his 2001 article Savulescu upholds the following position: 'The reason is that it is bad that blind and deaf children are born when sighted and hearing children could have been born in their place' (Savulescu 2001, 423). In his 2009 article he developed his views further concerning disabilities: 'In this final section we shall argue that PB provides a better approach to the question of disability than the competing procreative principles' (Savulescu/ Kahane 2009, 284). According to Savulescu and Kahane, it needs to be stressed that 'disability is a context and person-relative concept. What may make it harder to lead a good life in one circumstance may make it easier in another' (Savulescu/Kahane 2009, 286). As a consequence of Savulescu's altered approach concerning disability, Savulescu and Kahane stress that 'on our account of disability, people do have reasons not to have a future child who is likely to be disabled if they have the option of choosing another who is expected to have less or no disability, although whether it would be wrong to do so would depend on the overall balance of moral reasons' (Savulescu/Kahane 2009, 286). They even reach the following conclusion, given their new approach:

> If a case can be made that deafness is not a disability in our sense – if it can be shown that deafness does not reduce well-being, or at least that in a given context deafness is not expected to be a disability, then PB would not give any moral reason not to select deafness. (Savulescu/ Kahane 2009, 289)

Their final remarks concerning disability are actually quite interesting and do have some plausibility. However, I wonder which of Savulescu's remarks represent his theory of the good life upon which his principle of PB is based. In his writings, it is possible to find a wide range of affirmative statements concerning the good life which are mutually incompatible. It seems as if he has a theory of the good for every difficult question from his critics. On the one hand, he upholds a perfectionist theory of the good life which can be identified with being more intelligent, healthier, having a stronger memory, and so on. On the other hand, he refers to what we regard as the good life, which can be named as a common-sense approach to the good life. Yet I doubt that it would be a perfectionist theory of the good, according to which it would be best if one quality were 'perfected' (highest possible IQ), which would be upheld if large groups of Western citizens were asked what a good life is. My assumption gets support from the fact that many US mothers who ordered sperm were interested in sperm from good-looking, and athletic Ivy League students rather than in the sperm of Noble Prize winners (Caplan 2012, 156). This insight provides some evidence in favour of the Renaissance genius as a widespread ideal of the good. Still, Savulescu might reply that this argument does not go against his theory. Intelligence still plays a central role concerning the choices of these US mothers, as they desire to have an Ivy League student as sperm donor. Finally, his statement that PB is neutral to classical philosophical theories of the good life has to be mentioned, too. 'PB is neutral with respect to such philosophical disputes about the nature of the good life' (Savulescu/Kahane 2009, 279). However, it could be argued that the PB can be combined with various theories of the good, as it is logically separated from theories of the good. Savulescu combines it with his common-sense approach, due to its being the appropriate universally valid concept. Hence, his common-sense approach focuses primarily on the following qualities as necessary constituents of a good life:

1. not being disabled, whereby disability is seen as a context-dependent quality;
2. no disposition for mental illnesses;
3. having good health;
4. good capacities to concentrate, memorize, and being emphatic;
5. high intelligence.

Points 1 to 3 can be summarized under the category health, and 4 and 5 under the category cognitive capacities. Hence, the common-sense approach differs from the Renaissance genius concept in so far as the Renaissance ideal includes the possession of strong bodily capacities like being forceful, beautiful, and sporty, which is not demanded by the common-sense approach of the good. From my perspective, it makes sense to exclude bodily capacities like being forceful, beautiful, and sporty from the list of necessary qualities for living a good life, because there are several intuitively plausible reasons for holding that the aforementioned bodily capacities are not necessary for living a good life; for example Serge Gainsbourg seems to have had quite a fulfilled life without having had exceptional bodily capacities (at least from the perspective of a classic or Renaissance ideal). I will provide some philosophical reasons for this position in the next section.

4.2.3 Radical pluralism

Many people might respond to the initial two suggestions that either the first or the second is not so bad, and maybe even plausible. However, in the previous section I have already mentioned some reasons why a strengthening of the cognitive capacities might not always be in everyone's best interest. The remaining issue in question concerning the common-sense suggestion is that of health. This argument also reveals why philosophical doubts can be raised concerning the Renaissance ideal suggestion. Even Habermas seems to accept health as an all-purpose good and consequently allows genetic interventions in order to promote health (Habermas 2001, 48, 92). Yet, thinkers like Habermas and Savulescu do not sufficiently recognize the philosophical demand of what it means to claim that a judgement is universally valid, which implies that it is actually the case for all entities that belong to the human species, at all times, independent of their personal and cultural circumstances and psychophysiological demands. These psychophysiological demands are responsible for what is needed for someone to live a good life, and these demands differ radically from person to person and from time to time, because they change during the various stages of life. I am talking about psychophysiological demands in order to stress the intimate connection between these two faculties which for a long time were culturally separated from one another by being referred to as immaterial soul and material body, which is a highly implausible account of our human constitution. A non-dualist, relationalist account of what it means to be human is a far more promising way for describing ourselves – or should I say for what it means to be a metahuman, given that the concept of the human for many of us today still implies the concepts of an immaterial soul and a material body.

Given these psychophysiological demands, human choices which differ from those made by the majority of people should not be taken lightly. It is far more likely that the choices of the majority of people are determined by Heidegger's 'man'. They act according to the rules 'one' is expected to act upon. Only by listening to and acting in accord with someone's psychophysiological demands does a person become authentic. Consequently, the following acts can be understood as being based upon authentic wishes. This does not have to be the case in all instances, but it can be the case: (1) person A wishing to die; (2) person B desiring to have her healthy leg removed; (3) person C wishing to eat parts of himself; (4) person D not wanting to be cured of her manic depression; (5) person E regarding his deafness as an advantage but not as a disability. The list of examples could easily be continued. If the Renaissance genius account or the common-sense account of the good were universally valid, these wishes could not be accepted as authentic ones, but would have to be seen as expressions of a sick mind. I do not think that this has to be the case. By claiming that these wishes represent insane states of the mind, these persons are treated paternalistically and in a violent manner, as their wishes are not recognized as theirs, but is claimed by others that they know better than oneself what is in one's own interest. I regard such a way of treating people as highly problematic, because the otherness of someone else's wishes does not get appreciated appropriately. Initially it was difficult for me to imagine for myself that a deaf person is not disabled, but merely different. However, by recognizing the wide range of preferences, choices, tastes, and cultures in all parts of the world at various times I realized how important it is to recognize that a different human being might regard different capacities as important and different shapes as attractive.

In the fields of artistic appreciation, sexual preferences, and preferred tastes in food and drink, it becomes particularly clear that individual preference differs from time to time and from person to person. It is even the case that one individual's set of preferences is usually not matched by anyone else's set of preferences. Person A may hate spinach, white bread, beer, and honey, while B cannot eat spinach, white bread, beer, and sugar, and C may reject consuming cucumbers, white bread, wine, and honey. Any attempt to give a universally valid non-formal account of the good seems to be bound to fail, because there always seems to be someone who can convincingly claim that her authentic wish does not correspond to the concept in question. To acknowledge, recognize, and accept such a position means to affirm a radical pluralist account of the good.

It needs to be added that I am not claiming that it is always morally and legally acceptable to act in accord with one's own authentic wish. If you are sexually gratified only by raping someone, then your concept of the good is necessarily in conflict with what is legally acceptable in most countries today, and it is good that this is the case, because in this case your own concept of the good is (most probably) in conflict with the concept of the good of the person you wish to rape. A person's freedom ends where the freedom of another person begins. I will analyse this in more detail in the next section.

Furthermore, another remark needs to be made, because a radically pluralist concept of the good could be criticized as not being a proper transhumanist concept of the good, given that transhumanism aims for an increased likelihood of the coming about of the (further developed) posthuman. However, this argument is flawed, as there is no necessary conflict between this concept of the good and the potential to increase the likelihood of the coming about of the posthuman. Some of Darwin's basic principles must not be forgotten. The survival of the fittest does not imply that the strongest, tallest, or blondest human being survives. However, the fittest is always the one who is adapted best to the environmental and cultural circumstances. It cannot be excluded that there will be a situation in the future in which there will be so much loud noise that only the deaf will be able to bear the noise and survive. It seems to be impossible for us to know which capacities will be the ones that will guarantee our own survival and our living well in the future. We can decide only based upon our past experiences which capacities we regard as being in our best interest. I am not denying either that the Renaissance genius and the common-sense concept have a far-reaching appeal in this context. It is also the case that psychological and other empirical studies confirm that these concepts in question seem to be shared and affirmed by a great percentage of human beings (Bostrom 2009, 113–116). However, it is important for me to stress that it is inappropriate for both concepts to claim universal validity. Furthermore, it is not guaranteed either that the capacities affirmed by the initial two concepts of the good will be the ones which promote our fitness and our being best adapted to future environmental and cultural circumstances. Hence, there is the need to argue for the importance of the radically pluralist concept of the good life, which is closely connected to and based in the plurality of all of our psychophysiological demands, needs, and desires. Besides the philosophical reasons mentioned here in favour of the radically pluralist account, it could also be in the interest of a society to acknowledge the relevance and plausibility of a radically pluralist account: 'Don't put all you eggs in one basket'. As it is uncertain which qualities will promote our future survival, it is likely that an affirmation of plurality will increase the chances of human survival. Hence, both philosophical and social reasons support the radically pluralist position for which I argue.

4.2.4 Conclusion

The concepts of superman, Nietzschean overhumans, or human perfection are regularly tackled when dealing with transhumanism. The various meanings of perfection and differences between what is being upheld by transhumanists are usually not analysed in detail. In this chapter, I have given a summary of three concepts of the good which are associated with transhumanism. The first two might better represent traditional varieties of transhumanism, whereas the final one, the one in favour of which I argued, might not represent mainstream transhumanism, even though there are many areas and transhumanist protagonists whose suggestions bear a deep intellectual affinity to the concept of the good which I am suggesting, as the concept of morphological freedom is a central one in transhumanism.[10]

In any case, I wish to stress that I am arguing in favour of a weak version of both trans- and posthumanism so as to provide an approach that lies between these two movements. Consequently, it can be referred to as metahumanism. It lies between trans- and posthumanism, but it also places itself beyond a dualist understanding of humanism. *Meta* means both between as well as beyond. Thereby, metahumanism, which belongs to the neo-Nietzschean and neo-Heraclitean traditions, represents an inclusive alternative to the other two movements (del Val/Sorgner, 2011, 1–4).

In the first part of this section, I described how the notion of a weak transhumanism can be understood. Let me add a short note concerning the concept of a weak posthumanism. Posthumanists are concerned with non-dualist thinking and acting. A strong version of posthumanism assumes that human beings are merely gradually different from other natural beings, and it is this insight which ought to be legally obligatory. A weak version of posthumanism, on the other, might also affirm that human beings are merely gradually different from other natural beings, and this insight ought to be a legitimate perspective, that is, it should not be legally impossible to think and act accordingly (Sorgner 2013a, 2013b).

By integrating and further developing central concepts of both contemporary movements, a metahumanist approach was developed. (Sampanikou/Stasienko eds. 2021, section D) It is based on the insight that both movements do not differ from each other as strongly as many of their proponents assume. Central premises of both approaches are commonly shared. By weakening the central claims of both movements, potentially violent implicit demands are also reduced. Thereby, the option of human beings leading good, flourishing lives according to their own idiosyncratic standards is promoted.

By stressing the problems related to the claim of putting forward a universally valid non-formal concept of the good, several reasons turn up which underline the relevance of a radically pluralist account of the good. Thereby I am not denying that, for instance, an increase in the health span is relevant for a great percentage of human beings, which again ought to provide governments with a reason for promoting research in this field. At the same time, without contradicting myself, I stress that even an increased health span does not necessarily belong to the concept of the good of all human beings. It is an enormous cultural achievement to recognize and allow the radical plurality of the good. By trying to move beyond a non-formal account of the good, individuals are treated in a violent and paternalistic manner. Luckily, we have already moved far beyond the cultural situations in which religious or aristocratic leaders had the right to decide upon the concept of the good according to which their subordinates had to live. This does not mean that we are already living in a cultural and social situation in which the need to criticize legal and political structures no longer exists. A violent and morally problematic type of paternalism is still a widespread phenomenon. It can be found in cultures all over the world. Once I realized what an enormous achievement it has been to be able to live freely, in accordance with your own needs, desires, instincts, and wishes, and to be able to freely express your worries, criticisms, and opinions, I also became conscious of promoting this achievement of making people aware of what has been gained as a consequence of the Enlightenment processes which have taken place. Again, it is not the case that I can claim that we are living in paradise. There are many highly problematic issues still to be found even in countries like the Netherlands, Switzerland, or Germany. By becoming aware of the relevance of a radically pluralist concept of the good, it also becomes easier to detect potentially problematic paternalistic structures, so that it becomes possible to share one's insights with others.

In Germany the legal aspect of biotechnologies, reproductive medicine, and medical beginning-of-life issues is particularly problematic. It is the result of German history, in particular the experience of the 'Third Reich'. However, by trying to avoid the rise of totalitarian structures, the danger arises, too, that other totalitarian structures will come about as part of this process. A good example is Singer's position concerning personhood, towards which German academics and the educated public reacted with great outrage. It was their goal not to allow such a view to be expressed in German universities and in public discussions, as it was regarded as one which relativizes the value of human lives. Relativizing the importance of human lives also occurred in fascist Germany. This was the main reason for the strong reaction to Singer's view. However, thereby Germany reinstated another totalitarian structure. It limited freedom of expression, which was also decreed by the political leaders of the 'Third Reich' (Singer 2011b). In trying to avoid totalitarian structures, such structures were renewed. Being aware that totalitarian structures are in conflict with the plurality of human living and of human concepts of the good might be helpful for becoming aware when such structures do come about. In these cases, an analysis of whether the structures in question are problematic or not needs to be initiated. In some cases they are necessary, for example to legally forbid rape. In many cases, however, they are morally problematic and merely represent an attempt to dominate a certain interest group. There can be an attempt by the majority to dominate a minority via a widespread prejudice, or an attempt by the ruling class to dominate subordinates, as can still be the case, that political leaders make decisions which are not backed up by the majority of those who voted for them. By realizing, acknowledging, and living in accordance with the insight of a radical plurality of the good, violence against individuals is reduced and human flourishing is promoted. The earlier reflections merely provide an initial hint concerning the relevance of this concept. However, the question concerning the good and human perfection is of central relevance for all other philosophical, ethical, cultural, and artistic fields, topics, and domains, and ought to be analysed and considered appropriately.

The reflections on the good life also have implications concerning the question of how to plan the future. Should we have utopias or visions? Which types of planning, political, and social structures are needed? Often transhumanist ideas are connected with utopias, where concepts of the good a realized in a perfect future social organization. Huxley's *Brave New World* is probably the best-known example of such a future society in which gene technologies are used. However, before, I tackle the issue of utopias, I will present some further reflections on the issue of what harming a person should mean.

4.3 What does it mean to harm a person?[11]

The central task of my reflections in this section is to deal with the question of what harming a person could mean. In the first part I will be critically concerned with current ways of dealing with the concept of human dignity, and why it is no longer plausible to uphold these regulations, whereby I will particularly focus on the legal dimension. It will emerge that the person–object dichotomy can simply be upheld in its traditional manner. Given the plausibility of these reflections, the legal implication that it ought to be legally prohibited to treat another person merely as a means becomes implausible, too, as it rests on the aforementioned distinction between persons and objects. As an alternative, I will suggest that, legally, we ought to separate ontological insights clearly from normative ones. However, this does not mean that the same ought also to apply to moral reflections. In the following part I will focus on the moral status of personhood, and develop some philosophical reflections which could be upheld on the basis of a posthuman paradigm shift. Thereby I will develop a gradual concept of personhood which implies several levels of personhood depending on the corresponding capacity for suffering, which needs to be established empirically. Once the direction of an alternative concept of personhood becomes clearer, which implies that some animals become a person and which provides some reasons for holding that some sufficiently developed embodied AIs could also deserve personhood, the question of what it could mean to harm a person can be reflected upon further. However, the issue of harming is so complex that I plan merely to raise some questions which need to be confronted when clarifying the concept of harming. I do not attempt to propose a definite concept of harming, but stress that the concept is closely related to cultural circumstances, so that there is a continual need to engage with the concept, redefine it, and adapt it to the current cultural situation. Not proposing a clear concept of harm is not a shortcoming but, rather, a necessity.

4.3.1 Human dignity[12]

The concept of human dignity is at the core of many international constitutions. It plays a particularly central role in the German basic law (Sorgner 2010a, 23–29). Kant's concept of dignity is particularly influential in this context (Sorgner 2010a, 82–108). There are two ethical aspects which German law inherited from Kant, both of which are highly problematic. Firstly, the German constitution acknowledges that animals are not objects; nonetheless they are supposed to be treated like objects. Hence, the categorical dualistic separation of animals from human beings is implicitly contained in this judgement. Secondly, it is legally forbidden to treat a person merely as a means. This insight applies both to oneself as well as to other persons, which becomes clear in the following two regulations. Firstly, peep-shows in Germany are legally forbidden, even if it is the case that it is the dancer's autonomous wish to earn money in that way (Welti 2005, 397). Secondly, it is legally forbidden in Germany to shoot down a hijacked airplane, even though it appears to be flying directly into a nuclear power station, so long as there are innocent persons on board (BVerfG, 1 BvR357/05 from the 15.2.2006). In both cases, the regulation was justified by reference to the Kantian thought that it is morally false to treat a person merely as a means. In the following reflections, firstly, I will question a basic assumption on which this regulation rests, secondly, I will consider which options follow from these reflections, and thirdly I will analyse the challenges related to the options mentioned. Thereby it should become clear that the German law concerning the moral prohibition of treating a person solely as a means needs to be altered.

4.3.1.1 Challenging Kant's basic assumptions

The Kantian moral prohibition of treating a person merely as a means rests on the distinction between persons and things. Persons participate in the world which is governed by the laws of nature and the laws of freedom. Things, however, participate solely in the world which is governed by the laws of nature. This distinction implies that only persons do not solely belong to the natural world (Kant, 1902ff, vol 4, 428–434). Kant did not affirm an anthropocentric conception of personhood, but a logocentric position of personhood, as he did not hold that only human beings are persons. In the German legal context, however, the distinction between persons and things turns into an anthropocentric conception, as only human beings are viewed and legally treated as persons. Is this a plausible anthropology today? Darwin, Nietzsche, and contemporary posthuman thinkers might have reasons for doubting this conception (Badmington 2000, 9). Given recent biological research, given that human beings and the great apes have common ancestors, and given that a basically naturalist understanding of the world applies, it is more plausible to hold that there is merely a gradual difference between human beings, great apes, plants, and maybe even stones. Nietzsche's anthropology provides a possible non-dualist anthropology, which attempts to grasp the related concepts philosophically. Thereby, all entities turn into constellations of power-quanta and organisms and human beings are seen as a specific type of animal, sometimes even a 'sick animal'

(KSA, GM, 5, 367). However, this sickness, in Nietzsche's thought, is not necessarily a deprecation of human beings. The aspect of sickness has several implications. It means that they have developed a special capacity, namely the capacity of not immediately having to follow their instincts. This can have both beneficial as well as problematic consequences. It can be beneficial, as it enables human beings to create culture, develop technologies, and realize sublimation processes. It can be problematic, as it separates acts from the immediate realization of instincts, whereby instincts in many cases are more reliable concerning one's own interest than are intellectual reflections (Sorgner 2010a, 184–191).

4.3.1.2 Moving beyond Kant's basic assumptions

As both philosophical and scientific reflections have led many people today to doubt the Kantian anthropology on which the German basic law rests, what can be done to take these insights into consideration? It needs to be stressed that it is problematic to refer to the posthuman insights as those that are generally accepted. This is not the case. In Germany there are many citizens who still uphold a traditional Christian understanding of the world, which affirms the basic assumptions of Kantian reflections. Still, it needs to be asked whether a social-liberal democracy ought to be based upon a premise that affirms a strong metaphysical anthropology, namely one which is most consistent with an anthropology that views only human beings as being constituted of a material body and an immaterial soul. Animals, plants, and stones, on the other hand, are regarded as objects and as not participating in any world outside the material naturalist one. This seems to go against the fundamental norm of negative freedom on which democracies rest. In the case of Germany, it can be said that about 30% of citizens can be classified as naturalists, sceptics, or atheists, who are being treated paternalistically by this type of legal regulation.[13] A problem related to this group of people is that they are not sufficiently politically organized to efficiently act against such regulations. Members of the Catholic and Protestant churches, on the other hand, have strong institutions and hence have an enormous amount of power to influence political decision-making processes. However, the fact that at least one third of the population is treated paternalistically, in an aggressive manner, by these types of regulations goes against the central value which the norm of freedom ought to have within a democracy. Therefore, these regulations ought to be revised. What does revising the basic law imply? The current implications of the basic law are such that it has strong metaphysical implications, such that only human beings are seen as participating both in a material as well as in an immaterial world, but it is problematic that the basic law of a liberal-democratic society has an ontological basis. It cannot be an appropriate way of reacting to these insights to simply replace this regulation by another one, such that human beings and animals are seen as merely gradually different. In this case, one ontology would simply be replaced by another one. Instead of such a substitution, it would be more appropriate to stress the norm of negative freedom whenever the ontological implications of anthropology become relevant. However, the main question that I intended to address here was the prohibition of treating a person merely as a means. So far, I have described that the Kantian moral prohibition implies the ontological distinction between persons and things. Persons have autonomy, and hence dignity, which implies that no finite value can be attributed to them. Things, on the one hand, have a finite amount of value, which is also the reason why they can be treated solely as a means. Persons, on the other hand, cannot be identified with a finite amount of value, and consequently must not be treated merely as a means. Hence, the intellectual basis on which Kant's moral prohibition rests is a highly problematic ontological understanding, which currently is not shared by at least one third of German citizens, as mentioned earlier. Still, they are forced to be judged on this basis, as this regulation is part of the German law. It has consequences such as the prohibition of peep-shows as well as the prohibition of shooting down hijacked airplanes with innocent persons on board.

If the prohibition of treating a person solely as a means rests on the aforementioned ontological basis, the question needs to be addressed as to the consequences it has concerning this prohibition, given that one merely views a gradual difference between human beings, animals, and other entities. Two immediate options come to mind: firstly, due to there being merely a gradual difference between human beings and other entities, there are no more things, and hence it will have to be morally prohibited to treat any entity merely as a means. Secondly, it can be argued that the prohibition of treating a person solely as a means does not even apply today as a universally valid regulation; for instance, if someone has committed murder, in specific circumstances he can be sentenced to death. This seems to imply that he is being treated merely as an object. Hence, treating a person solely as a means can be both morally legitimate as well as morally illegitimate (Hoerster 2013, 11–23). If this judgement applies to persons, then it applies also to all other entities, given that there is solely a gradual distinction between all entities in question.

4.3.1.3 Challenges related to these moves beyond Kant's basic assumptions

There were two major suggestions concerning these moves beyond Kant's basic assumptions. Firstly, the moral prohibition of treating persons solely as a means was reinterpreted such that all entities, from stones to great apes and human beings, become persons. Secondly, the moral prohibition mentioned was dissolved, as the distinction between morally legitimate ways of treating a person merely as a means and morally illegitimate ways of treating persons merely as a means was introduced. The second case seems to imply that the prohibition in question no longer applies to any case, such that the question has to be asked anew: what is moral and how can we conceptualize morality? The first reply, however, raises different questions, as the question arises of what it means and what are the implications of treating someone merely as a means. Does it imply that I must no longer eat tuna? Is it morally problematic to walk on grass?

A further issue has to be considered in this context, too, which has been mentioned earlier. If the moral prohibition is altered in one of these two ways, does this not imply that one ontology is simply replaced with another one within the legal context? Is it not problematic to have any ontology which influences legal decisions, as a social-liberal democracy must imply openness to a great variety of ontologies and must not judge its citizens on the basis of any ontology, due to the morally problematic implications involved? If this is indeed the case, then it might be advisable to move beyond any ontological discourse when dealing with any legal discourse, as this is the only way of remaining ontologically neutral such that no morally problematic intrusion of the state into the personal decisions of its citizens occurs.

In the preceding reflections, several challenges of legally dominant regulations have become clearer which have a particular relevance for the German legal context. Given these reflections, it seems appropriate and necessary to move beyond the prohibition of treating other persons solely as a means and also beyond the tradition of allowing ontological positions within the constitution of a social-liberal democracy, as both judgements contradict the initial premises of such constitutions and hence are self-contradictory. In any case, given the paradigm-shifting alterations which have occurred concerning our anthropological self-understanding, the concepts of who should count as a person from a non-logo- as well as non-anthropocentric moral perspective and what constitutes harming a person need to be reflected on anew. Firstly, I will deal with the notion of personhood, then with the notion of harm. Legally, it might be most in tune with a liberal and democratic constitution to separate ontological and normative perspectives, which implies the need to come to pragmatic normative decisions concerning ethical challenges. Even the legal prohibition to treat another person merely as a means can no longer be plausibly justified, due to the implausible ontological connotations of this demand. Morally, this is a separate issue. Here, we can ask anew which concept of the moral status can still be upheld on the basis of an ontology of continual becoming. Yet, it is no longer a foundationally justified concept of a moral status, but merely a fictive one. For pragmatic purposes, we attempt to create plausible criteria, knowing that these criteria are contingent nodal points. We agree upon them now, as they are widely shared, but they do not represent an ultimate foundation for a moral status. Given further information or a different *zeitgeist*, our judgements concerning what should count as a moral status will change. Then, our ethical reflections need to be adapted. It is an open question who should count as a person and what constitutes direct harm done to another person. Should indirect harm also be regarded as immoral? How far can direct be distinguished from indirect harm? Can there be immoral acts which do not involve any harm being done to a person?

The questions just listed have arisen as a consequence of the latest technological innovations. If we have a sex robot which acts as if it does not wish to be engaged in sexual acts, does this mean that you are raping the robot if you nonetheless try to and do have sex with the robot? Does not rape presuppose the absence of consent? How can a robot consent at all? Does not consent demand a certain type of competence which robots do not possess? One argument claims that fictively acting immorally, for example committing a make-belief rape of a robot, increases the likelihood of actually acting in a similar fashion. Yet, is this the case at all? Empirical studies on computer games seem to suggest that there is a difference between fictive killing in a game and the likelihood of actually killing another person.[14] It can still be argued that killing is different from having sex, as the latter is more directly connected to one's more important desires. Whether or not this is the case needs to be further studied empirically. So far, there is no clear evidence for making such a rigid distinction between make-believe killing and make-believe sex. In addition, there are further arguments which suggest that acts can be immoral even though they do not directly cause harm to anyone; for example it is sufficient for an act to be considered immoral if that act normalizes a procedure which ought not to be normalized. Does the make-believe rape of a sex robot normalize the act of rape?

I think it is important to distinguish between significant and insignificant distinctions. In the case of a make-believe rape of a sex robot no direct harm is being done to another person. However, it can be objected that acting thus is morally wrong because it normalizes an act which we do not wish to see becoming normalized, and that, due to an altered social meaning of rape, the likelihood of rape occurring may be increased. The implicit logic is that a make-believe rape of a sex robot is morally wrong because it indirectly harms other persons due to the normalization of rape which goes along with it, or because it presents an immoral act, that is, rape, as morally legitimate. However, neither of the reasons seems to correspond to the issue in question. Rape does not become normalized. Having sex with an entity which does not suffer, and which has neither consciousness nor self-consciousness is normalized. If two competent people engage in a mutually agreed upon activity, no harm is involved. Parallels between types of events may be drawn only if the events are structurally analogous in a morally significant manner. This is not the case when we compare sexual acts between a competent adult and a sex robot with the sexual act between two competent adults, whereby one of the adults does not consent to the act.

This is also the reason why incest between consenting adults is morally legitimate on the basis of a liberal ethics of autonomy. In this context, the potential consequences are often referred to, as the likelihood of having a child with a handicap is increased in the case of an incestuous sexual relationship. However, is this a solid argument? If you start from a liberal ethics of autonomy, which is the ethics I subscribe to, it is not, as no person is harmed. Would it be different if you held a utilitarian perspective? Even given a utilitarian perspective, it is not a straightforward argument. Bringing another human being into existence is usually in the interest of the human in question, no matter whether the child is handicapped or not. If you hold that it is not worth living if one is handicapped, one could argue that an incestuous relationship between consenting adults is immoral. However, the claim on which this argument rests upon is highly implausible, and not one which I share.

Given the plausibility of what I have just argued, it is implausible to claim that there can be immoral acts which do not directly harm another person. However, it still remains an open question as to how we should plausibly conceptualize a person, and what constitutes the direct harm being done to another person. Both questions will be dealt with in the following sections.

4.3.2 Twisting the person–thing distinction

In the humanistic tradition, it used to be upheld that only humans count as persons. This is slightly exaggerated, as angels and God were also often regarded as persons. The point is that the only living beings in the natural world who were regarded as persons were human beings. This is still the dominant outlook in monotheistic religions, and also in most constitutions in all parts of the world. After Darwin, however, this view is no longer plausible. Respect for entities, that is, personhood, should depend upon morally relevant capacities, and not upon the former speciesist position. Fertilized human eggs without a nervous system and without a brain should not have a higher moral status than adult chimpanzees, who recognize themselves in the mirror, as entities which have the capacity to suffer seem to deserve special moral consideration. Yet there are differences with respect to the capacity of suffering, depending on the qualities which entities possess. Peter Singer ranks entities which can suffer strongly, because they have self-consciousness (have a temporal awareness of past, present, and future; that is, the ones that pass the mirror test), over those which have only consciousness and sentience (and live merely in the present moment). In this case, too, AIs with sensors, that is, embodied AIs, would necessarily be excluded from a moral status. Maybe this alternative view is also implausible. Is sentience necessary for personhood? There are humans who cannot feel physiological pain. Should they not count as persons? Cognition might not be dependent on consciousness either, as there are indications for the possibility of non-conscious cognition (Hayles 2017). Maybe cognition can also lead to a type of cognitive suffering, which embodied AIs could also realize. How could we compare cognitive, emotional, and physiological suffering? Is there the possibility of developing an empirical way of investigating the intensity of suffering and of comparing different types of suffering?

The Earth came into existence about 4,5 billion years ago. But whether life was established on Earth or on another celestial body and afterwards reached Earth is an open question. The problem is that there was only inanimate matter at first (water, rock, gases), and then suddenly a moment came when life arose, together with entities capable of self-movement. Researchers believe that life came about 3,5 billion years ago in the sea. At first everything was determined solely by causal processes. Suddenly, however, certain entities developed a principle of order by means of which they could develop. Scientists often speak of a primordial environment populated by water, gases, and electric lightning, which gave rise to the simplest forms of life. Six million years ago, the apes living today and *Homo sapiens sapiens* still had common ancestors. Two hundred thousand years ago, *Homo sapiens* came about. It would be naïve to assume that *Homo sapiens* will still exist in six million years' time. Species must adapt in a constant way.

In ancient philosophy all living entities were described as animated. In Plato's work, the soul was not yet identified exclusively with the soul of reason; animals and plants were also given certain types of souls. Everything that could move by itself had a soul and was thus alive. However, how life could arise from (apparently) inanimate matter is astonishing – just as the fact that something is there and not nothing is confusing. Furthermore, the situation became more complicated, as these simple life forms have evolved and become more complex. Some of them even developed a special ability for emotions, perceptions, abstraction, spoken language, written language, artistic and numerical forms of communication, reason, and creativity. How could such abilities develop from water, electricity, and gases? In addition, awareness has developed in the course of these processes, that is, the ability to feel pain and to perceive and react to the environment. Finally, living beings have developed as beings who had an understanding of time, of the past and the future, and who recognize themselves in the mirror, that is, possess self-consciousness. At present, in addition to humans, nine other animal species are known (great apes (Orangutan, chimpanzee, bonobo, gorilla), dolphins, elephants, orcas, the Eurasian magpie, the cleaner wrasse), of which selected members have passed the mirror test. How could this develop on the basis of water, gases, and electricity alone?

Something must have been added from the outside, such as a divine spark: the philosophers of Western cultural history (decisively influenced by Plato and then developed further through the Stoics, Descartes, and Kant) predominantly argued that both the immaterial spirit and the gift of free will would be able to explain these abilities. They usually adhered to the primacy of a non-empirical understanding of reason. This idea is still anchored and valid in German law today. Animals are not things but should legally be treated as things. Only human beings possess dignity (Sorgner 2010a).

These considerations have been shaping our thoughts, actions, and the culture of the Western world since Plato. However, some of these considerations exist only as relics and encrusted structures of traditional considerations. This tradition began to crumble several hundred years ago. When Darwin formulated the theory of evolution and Nietzsche proclaimed the death of God, the cultural movement away from dualistic ontological thinking gained enormous importance. Instead of assuming that something would be added from the outside in the processes described earlier, it is now assumed that all these developments have happened on their own as emergent processes, without any magical connections between material and non-empirically accessible entities. Transhumanism also rose from the tradition of evolutionary and naturalistic thinking. This is an important reason why the conservative American political scientist and cultural critic Francis Fukuyama calls transhumanism the 'most dangerous idea in the world' (Fukuyama 2004, 42–43). The term transhumanism was coined in the context of the emergence of evolutionary thought. It was described by Julian Huxley in an article published in 1951. I think his definition is still correct. Furthermore, Huxley also coined the term 'evolutionary humanism'. The relationship between evolutionary humanism and transhumanism represented today by the Giordano Bruno Foundation needs to be clarified further (Sorgner 2016e). Huxley was also the director of the British Eugenics Society.

The close connection between transhumanism and genetic engineering also exists in contemporary transhumanism. Furthermore, the focus on cyborg techniques and AI has been added, because with the help of all these techniques the previous limits of our humanity can be transcended. This objective is relevant for many reasons. The central assumption is that the probability of leading a good life is increased if we develop further through training, practice, and also with the help of other techniques. Furthermore, this approach is important, since all life is always threatened by the possibility of extinction. Life has evolved. Just as dinosaurs became extinct, so could human beings become. Extinction or survival depends on how well we are adapted to our environment. However, as our environment is constantly changing, there is a perpetual need for us to change too. If we develop techniques which can help us, we can increase the probability of survival.

The two most promising techniques are genetic engineering and AI (Sorgner 2018b). With the help of different genetic techniques, especially CRISPR/Cas9, people could develop in a carbon-based way, that is, people could become organic trans- or posthumans, whereby the posthuman either still belongs to the human species but has at least one characteristic which goes beyond the limits of current human capacities, or can become a representative of a new species. The possible alteration processes using cyborg techniques and AI are even more radical. They involve an intensified fusion of brain–computer interfaces until the development has gone so far that mind uploading becomes an option and we can store our personality on a hard disk, which, according to the Google futurist Ray Kurzweil, should already be possible in a few decades (Kurzweil 2006; Kurzweil/Grossman 2011). In this case, the posthuman would no longer be a carbon-based being but a silicon-based one. The fact that a transmission of our personality is conceivable, at least in principle, becomes clear simply by the fact that all the cells of our body renew themselves every seven years. Nevertheless, we remain a continuous unit. However, whether what exists on a carbon basis can be transferred to a silicon basis is quite questionable (Sorgner 2018b). At present we do not know of a silicon-based life form; or should self-replicating computer viruses already count as a kind of life (Schrauwers/Poolman 2013, 1–54)?

With this development into a silicon-based posthuman, we are facing new ethical challenges. The dominant way of assessing the moral status of entities (which is currently widely held in all parts of the world) is based on the cultural tradition which assumes that the evolutionary development into human beings is connected to the rather decisive event of the incarnation, something that goes beyond the naturalistically explainable evolutionary process. At this point, God's divine immaterial spark, our reason, entered into us and connected with us. This process is responsible for the fact that only we humans possess something that goes beyond the purely natural world, which is why only humans have subject status. All human beings are subjects, persons, and bearers of dignity. All other entities are objects, non-persons, and can therefore be traded, as they possess a finite value. This categorization is both morally and legally dominant. There are only a few legal exceptions, for example when a court in Argentina granted the status of a person to a great ape.[15]

A particular legal situation appears in Germany. People have dignity; animals are not things, but should be legally treated as things. This assessment which clarifies the idea that only human beings were given the immaterial divine spark is still legally valid. Or at least this understanding is suggested by the legal assessment (Sorgner 2010a).

This basic attitude is problematic in many respects. Every third German sees herself as a sceptic, naturalist, or atheist, which implies that the legally suggested understanding of animals is not shared by them. Legislation thus patronizes the majority of Germans in a morally problematic and paternalistic way. Such a situation is unacceptable in a liberal-democratic state. In my opinion, therefore, ontological implications should also be banished from the constitution and replaced by contingent-normative attitudes in order to ensure that a plurality of ontologies is lived out.

However, the question of the moral status of apes and other animals is yet to be adequately addressed. The strongest intellectual counterproposal comes from Peter Singer, who makes the following considerations regarding the concept of human dignity (Singer 2011a, 76–77). He labels the traditional humanist ethical theory as speciesist. Human life is preferred in a morally illegitimate way, that is, it is preferred, even if it has no morally relevant qualities. In this context, a human embryo has human dignity, even if it does not even have a nervous system or brain and otherwise does not fulfil the prerequisites for perceiving pain. However, the property of perceiving pain should, in his view, be the basis for giving a being an elevated moral status, and the more intensively someone can feel pain, the higher the corresponding moral status should be.

The suggested ability to suffer seems to be a plausible criterion. It needs to be kept in mind that this criterion is not an essentialist and ultimate one, as was the case with the divine spark. It is a merely a widely shared judgement. The ability to suffer is a contingent criterion. However, many people find it convincing that the minimization of suffering is desirable. Hence, the posthumanist counter-argument that by introducing the moral relevance of suffering one affirms a new essentialism does not stand up. It is not a new essentialist philosophical insight that suffering matters, but merely a currently widely shared contingent insight. A second posthumanist counter-argument also needs to be refuted, that is, that by introducing suffering as a morally relevant criterion one is affirming a new dualistic categorization of life, that is, the good against the others, the bad, the monsters. This is not the case, as there are many different degrees of capacity for suffering. Perhaps these can even be determined empirically in the future. If we managed to establish different degrees of suffering, we would end up having a wide continuity of concepts of personhood. Thirdly, many critical posthumanists accuse transhumanists to be anthropocentric. This is correct for some transhumanists. However, many transhumanists stress the need to also apply personhood to non-human animals. This alone makes it clear that these transhumanists do not affirm a human-centred, but rather a person-centred stance whereby personhood should depend on the capacity of experiencing suffering. Many animals can suffer, but human embryos without a brain and nervous system do not possess this ability. Consequently, some humans beings would no longer be persons, but some animals would have to be regarded as persons, and further gradual distinctions of various degrees of personhood would have to be introduced, too. A final comment regarding posthumanist critiques of this suggestion concerns the fact that the concept creates new hierarchies, but (at least some) critical posthumanists wish to develop a non-hierarchical ethics. Does the aforementioned suggestion lead to a new hierarchy? This is indeed the case, but a non-hierarchical world cannot even be conceptualized convincingly in theory. Every decision means that one action gets preferred over many other options for action. The demand of a non-hierarchical society is necessarily a performative self-contradiction. So, we have established that a posthumanist critique of a suffering-based ethics lacks a plausible line of argument. Yet what are the implications concerning personhood, if we regard suffering as morally relevant?

Out of these considerations Singer develops a counterproposal. Someone who has only consciousness lives exclusively in the present. But if a being has self-consciousness, then it perceives itself as a continuous entity, an entity that existed in the past, exists now, and will probably still exist in the future. If a being has sentience and self-consciousness,[16] then the ability to feel pain should be much higher than in beings who have consciousness only and live in the moment. At present, only a few members of nine animal species, apart from humans, have this ability.[17] It should also be noted here that only about 65% of two-year-old humans have the ability to recognize themselves in the mirror, which is the most important test of self-consciousness (Amsterdam 1972, 297–305).

However, it may well be asked whether the mirror test is an appropriate test of self-consciousness. Dogs do not pass it. Could this be because the test is biased in terms of the ability to see? Could it not be the case that for some living beings the ability to hear or smell is more pronounced and linked to self-consciousness (Cazzola Gatti 2015, 232–240)? In any case, the consequences of Singer's ethics are enormous. If the choice were to save an elephant or a new-born with severe mental disability, then his theory would give a clear answer as to which was the morally appropriate choice. His answer is not shared by many enlightened people, which is an important reason why I find his considerations problematic. However, his theoretical reflections are quite conclusive. So, how can it be explained that the intuitive reactions to the consequences of his ethics are negative?

One explanation could be that our emotions are strongly influenced by our cultural imprint, which is anthropocentric. It is also possible that our emotional response to his reflections will change if our culture becomes less anthropocentric. If this were the case, then our emotional reaction to the aforementioned case might correspond more closely to the consistent moral judgement Singer suggests, or a similar, even more plausible ethical, non-anthropocentric suggestion.

The crucial point why I am addressing this issue is that Singer's ethics do not seem to provide us with a well thought-out assessment of the moral status of AI or, for instance, of the humanoid Data from *Star Trek*. Let us assume that it would be possible to download the personality onto a hard disk, and self-consciousness could remain. Nevertheless, this being would not meet Singer's requirements for personal status, since (most likely) sentience would be lacking. At least at present, it is difficult to imagine that AI and robots possess emotions and can feel physical pain, since there is no organic body, which seems to be necessary for such sensations. Would a person uploaded to a hard disk lose their personal status through the process of mind uploading? It is highly speculative to reflect upon an uploaded posthuman, whom we may at best know from the film *Transcendence* starring Johnny Depp. But how should we generally deal morally with computers and AI? On the basis of anthropocentric human-dignity ethics, these entities would clearly be things that can be owned, destroyed, and sold. If we look at the relationship between soldiers and combat robots and between old people and care robots, which already exist to an increasing extent in East Asia, then this assessment seems too short sighted.[18] But on what philosophical basis could an assessment of the moral status take place? It could be argued that AIs can already pass the mirror test to some extent. Does this mean that robots with AI already have self-consciousness?

This example seems to indicate the limits of the mirror test. On the other hand, AIs with sensors pass the test only so far, if they are initially informed of what they look like. Living animals do not need this initial information. However, what we understand by self-consciousness could be nothing more than a special algorithm. In any case, it seems impossible to have physical pain without a carbon-based organism, and the ability to feel pain was decisive in giving a being a moral status, based on previous considerations. But perhaps these considerations are not sufficient.

Avishai Margalit emphasizes the relevance of dignity in not humiliating others (Margalit 2012, 150). What situation arises in case of humiliation? It is the situation of a relation in which one being puts itself above another and expresses contempt for the other. However, the humiliated person does not necessarily feel the physical pain that he would feel when he breaks his ribs. Rather, humiliation is primarily associated with the cognitive realization of not being appreciated. The process is also painful, but without having to be associated with physical pain. Perhaps it can be described as cognitive pain, which has to be linked only to cognition, but not to consciousness, if we can uncouple consciousness and pain. The possibility of decoupling seems to exist because there are signs that foetuses can feel pain without having consciousness; for example research indicates that from the sixth month onwards, foetuses have the physiological prerequisites for experiencing pain (Rollins/Rosen 2012, 466), but it is rather unlikely that we would attribute consciousness to foetuses at this stage. Wakefulness is a post-birth phenomenon. Is pain without consciousness possible? The relationship of the central nervous system, memory, brain, consciousness, and the ability to feel pain is extraordinarily exciting.

There is also evidence that cognition is possible without consciousness. One indication of this being the case are the results of a selective attention test, also known as the 'Invisible Gorilla'.[19] We are asked to count the number of passes of a basketball team and are then asked if we noticed anything special about the video. The kickboxing gorilla that walks through the playing field remains unnoticed by many. Nevertheless, it is in our cognitive field. Cognition and conscious perception are therefore two different phenomena. These considerations suggest that there is not only a conscious and an unconscious, but also that there may be a non-conscious cognition, as in the case of the invisible gorilla (Hayles 2017).

If both non-conscious cognition and the phenomenon of cognitive pain exist, as illustrated by the example of humiliation, then it may make sense to talk about the possibility of non-conscious cognitive pain in an AI with sensors, that is, an embodied AI, as no non-embodied AI could have access to a cognitive input. A further developed embodied AI might also cognitively realize that it is not appreciated, and in this case the cognitive pain associated with humiliation might be associated with this realization process. Data from *Star Trek* would have to be given a moral status against the background of these considerations. Whether Data ought to receive the person or the post-person status would have to depend, among other things, on the relationship between cognitive and organic pain. I think that an empirical means of analysing the intensity of various types of pain would be needed so as to be able to compare types of personhood.

The foregoing reflections show that Singer's suggestion that self-consciousness as well as sentience are necessary for personhood is not necessarily plausible, as there can be pain without consciousness. An additional challenge for Singer's concept is that there are human beings who cannot experience physiological pain.[20] They are not many, but as they would not be regarded as sentient it would follow that these human beings ought not to be regarded as persons, given Singer's reflections.

On the basis of Singer's approach, the hundred or so people worldwide who cannot feel pain would not even have the status of a person, as they lack sentience. However, they are also able to cognitively realize pain. They also often react analogously to other people, although they cannot feel physical pain. With injuries they learn to show similar reactions as are usual in most humans. They cognitively recognize the pain and act accordingly. It might be worthwhile to explore the relationship between cognitive, physical, and emotional pain together with them. It cannot be ruled out that the findings might also be transferred to AI. Given certain empirical data when experiencing cognitive pain, it might be possible to find a basis for comparing these findings with empirical data which could become available with embodied AIs. However, these further developed embodied AIs are not yet available.

In any case, we can conclude from the earlier reflections that neither sentience nor self-consciousness is necessary for being in pain. Hence, both elements which Singer regards as necessary for being a person fail as a plausible basis for personhood, even on the basis of his own suggestion that personhood ought to depend on the capacity of experiencing pain. Still, this most fundamental move of his applies. The concepts of personhood and being a human being are disentangled. There can be physiological, emotional, but also cognitive pain, and consciousness is not necessary for being in a state of pain. How to compare the intensity of the various types of pain remains an open question. I think that an experimental scientific basis would be needed to compare the various types of pain with one another. A separate article is needed to analyse these issues in more detail.

An additional question which could arise is whether the capacity of autonomy plays any role when it comes to the question concerning the moral status. If an increased autonomy is accompanied by an increased capacity for suffering, then we would be forced to answer this question in the affirmative. But is it the case that autonomy and the capacity for suffering are linked?

Autonomy is linked to reason. Reason has to do with the use of language, forming sentences and judgements, and making inferences. We do get to know things by means of inferences, and by combining our cognitive input with the capacity of making inferences. These are merely some initial reflections which hint at the possibility that an increased autonomy implies a higher degree of cognition, which again implies the possibility of cognitive suffering. An increase in autonomy towards hyperautonomy could be accompanied by an over-reason, and therefore a significant increase in the cognitive abilities of the hyperautonomous entity. This in turn would suggest that the possibility of the aforementioned cognitive suffering is also present in these entities. Whether a hyperautonomous entity should have the status of a person, a post-person, or no person at all depends on the relationship between cognitive pain and emotional pain. Is cognitive pain comparable to the ability of only conscious, self-conscious, or perhaps even meta-self-conscious beings to suffer? The android Data could be such a hyperautonomous posthuman. On the basis of empirical studies of humans who do not experience physiological pain, as well as highly developed embodied AI, efforts could be made to find criteria for the comparability of cognitive, physiological, as well as emotional pain. However, since we do not yet know of any digital entity capable of recognizing cognitive pain, further discussion of this question would be purely speculative. Further empirical studies with human beings who cannot feel physiological pain seem to be a good starting point for finding a scientific way of studying the intensity of pain, and for finding an empirical basis for making judgements concerning personhood. These reflections are far from being conclusive concerning the tricky issue of personhood. However, I think that these traces are worth pursuing further on our path towards a posthuman existence. It also has paradigm-shifting implications concerning a widely shared moral wrong, namely the moral demand to prohibit a person to treat another person merely as a means, which I dealt with earlier.

4.3.3 Reduce violence and avoid damage

After briefly addressing the question of a revised understanding of the moral status, some considerations need to be made to reduce violence and prevent harm. What is violence and harm? This is a complex issue. My main task is to indicate some reflections which ought to be taken into consideration, rather than a definite answer. Cultural conditions always need to be considered when specifying a convincing account of harm. Given the wide variety of thoughts which need to be taken into account, I suggest an 'as-good-as-it-gets ethics'. It is a contingent ethics, based upon the widely shared judgement that suffering is morally relevant. The higher an entity's capacity of suffering, the higher should be their moral status. The potentiality of an entity is not relevant, as it remains uncertain whether a potentiality will ever unfold, and a capacity becomes relevant only once it actually is morally relevant. A sperm does not possess morally relevant capacities. You and I do.

Any person should have the right of acting freely as long as no other person is harmed, as each person has idiosyncratic needs for realizing a fulfilled life. The right of acting freely includes many other rights such as morphological freedom, procreative freedom, or educational freedom. This judgement considers the relevance of acknowledging a permanently greater plurality of legitimate lifestyles, while at the same time it takes seriously the goal of reducing violence done against people.

These reflections do not imply the idea of a perfect utopia. Utopias are dangerous, because they sacrifice the present for an uncertain future. The best we can do is to take into account the given structures and current developments in order to find the best possible response to the challenges we face. Any other attempt will lead to the retotalization of society, which must be avoided. The past has taught us that the effects of totalitarian regimes must be avoided. The greatest efforts must therefore be made to create structures that make our system as liveable as possible. The best minds of our generation are needed to participate in the reorganization of society, so that the diversity of concepts of good can be further promoted, as can the consideration of the norm of negative freedom at a legal, social, and moral level. The following sections highlight selected problems related to the concern to avoid harm done to people. They also take into consideration some developments which are taking place at the moment, in particular the increase of digital surveillance, which I dealt with in detail in Chapter 2.

It is clear that the higher the digital monitoring rate, the more likely it is that you can be punished for what you have done, if it went against the law. But what should you be punished for? This issue becomes particularly problematic when it comes to religious needs, social cohesion, and the limits of autonomy. A striking example is male circumcision. It is a procedure in which part of the foreskin is removed, often for religious reasons. Experience shows that it also reduces the intensity of sexual stimulation of the circumcised.[21] However, in order to be a full member of Islam and Judaism, one must undergo this procedure. For the believer, this procedure is an enrichment, since it is a necessary element for initiation into a religious community. For a non-believer, it is unnecessary bodily harm directly done to a child. In Judaism, male circumcision should take place before the ninth day after birth. In Islam, on the other hand, it can occur up to the age of 13 years. This makes a difference, because human sensitivities have increased during this developmental process. Only 50% of children at the age of two years can recognize themselves in the mirror, which can be regarded as a sign of self-consciousness. From then on human children have an awareness of their own self which exists in the past, present, and future, in contrast to the time after birth, in which one exists mainly in the present. As a result of this changed self-perception, your ability to suffer increases. Therefore, the procedure is more severe for older children than for younger ones.

It has to be mentioned here that the meaning of restricting religious freedom also depends on the cultural circumstances of a country. The prohibition of male circumcision, as was considered in Iceland in 2018, has a different meaning there than in Germany, with its history of the 'Third Reich'. Legally undermining the possibility of Jewish religious practice in a country where the Holocaust took place has a different meaning than in a country like Iceland. The cultural aspect is therefore one that must also be taken into account when reflecting on the question of morally illegitimate forms of harm.

An additional issue concerns the issue of acts which a person ought to perform, and whether such legal duties could count as a morally illegitimate type of harm by a government against a person. In France there was an interesting case of a woman who was married to a Frenchman. She was supposed to become French in an official government ceremony, but for religious reasons she refused to shake the hand of the official who was responsible for granting French citizenship. In an immediate reaction, the official tore her paper in pieces, seeing a certain social cohesion as a necessary element of citizenship.[22] Should it be a legal obligation to shake the hand of a public official during such a ceremony, even if it violates a widespread religious obligation? Should you legally enforce handshakes in certain situations? Should you enforce a handshake if the person in such a situation has an infectious disease such as influenza? Is harm being done to a person by forcing them to shake hands with someone else, which they do not want to do for religious reasons? In how far has the coronavirus pandemic changed the cultural meaning of the handshake?

A related issue concerns an obligation to refrain from dressing in a specific manner. How is the Austrian ban on face veiling to be assessed? In Austria it is forbidden to cover one's face. A full ban on veiling has been introduced in Austria and in other European countries. This has consequences for people who wear costumes in public, and also for some female Muslims. Here, too, the question arises as to whether this ban is unnecessarily harmful by undermining the right of religious freedom. In this case, the question of personal identity also becomes relevant. If you drive a car, it must be possible to identify the driver; otherwise you cannot be punished for misconduct. This is usually done by taking a picture of your face if you are driving too fast. If you can be identified by an implanted chip, then prevention of identification by wearing a veil is no longer an issue. If face recognition is needed for identification purposes, there is a good reason for prohibiting car drivers from wearing a full-face veil for. However, even this regulation has significantly changed, due to the coronavirus pandemic. What used to be legally forbidden was changed into a legal duty.

These reflections show the complexity of clarifying the issue of avoiding harm and reducing violence. Such considerations also apply to Christian groups who wish to prohibit abortion. In Germany, most abortions are still illegal,[23] but if you follow a certain procedure you will not be punished for having an abortion. This means that fundamentalist Christians force other members of a social-liberal democratic and pluralistic society to act according to their worldview. This is clearly a case of unlawful harm, as in this case one ontology dominates another, which is not in tune with the basic pillars of a pluralistic democratic society. Such legal regulations can be explained by the effectiveness of encrusted paternalistic relics of our Judeo-Christian past, which are particularly strong in Germany. In Germany there are numerous morally problematic paternalistic structures that contradict the basic premises of a liberal-democratic society. Even though a large number of freedoms were established as a result of the Enlightenment, an enormous number of paternalistic structures are still dominant in many enlightened countries. These are the delicate issues that lead to morally problematic implications, especially within a society which embraces total digital surveillance more and more, such that the internet panopticon is realized.

Further significant challenges arise with regard to the question of universal public health insurance. Does it harm taxpayers if they are forced to financially support universal health insurance? This is not a stupid thought. If the norm of negative freedom is a central one, a good justification for such obligations must be provided. In the case of universal health insurance, it is possible to refer to psychological studies that show that an extremely high percentage of citizens associate a long health span with a better quality of life. If this is the case, then it can become a political obligation to promote the availability of universal public health insurance, even if a long health span is not associated by everyone with an increase in quality of life. However, since most people identify health promotion as an improvement in the quality of life, the introduction of universal health insurance can be justified.

The aforementioned cases of harm being done to a person are just a few concrete examples to illustrate selected challenges in reducing violence and harm being done directly to a person. There are many tricky cases, and each one must be carefully analysed to find an appropriate response. I also believe that cultural realities must be taken into account when dealing with such moral issues. Our cultural history has relevance for the way we deal with things in this day and age. All of our regulations are contingent nodal points, which need to be continually reassessed by a representative ethics council, which needs to be assembled carefully.

4.3.4 Conclusion

Who is a person and what counts as harming a person are questions which ethicists, philosophers, and theologians have dealt with for centuries. The latest cultural as well as technological developments imply new ethical challenges concerning this issue. At the same time, they also provide us with enormous amounts of benefits, which must never be forgotten. I particularly wish to stress hybridization processes which are currently occurring. Chips are finding their way into our bodies. I will take up the reflections from Chapter 2.

Human–machine interfaces, such as those developed by Elon Musk's Neuralink, are of particular current relevance. It is crucial that RFID chips are implanted in our bodies. In this way, the already existing human hybridization will be further advanced. Interaction with the Internet of Things will be made possible in this way. At the same time, an Internet of Bodily Things will be created, which can be helpful in promoting central human concerns. A widely shared desire is the promotion of the human health span. Rather than just talking about life span, it makes sense to emphasize the importance of health span, since for many people living in a state of constant suffering may not seem worth it. Psychological studies emphasize that many people identify the promotion of the health span with an instrumental or even intrinsic increase in quality of life.[24] If it is now the case that ageing processes are also widely identified as disease, then the political importance of promoting the health span is obvious. Human upgrading by means of RFID chips is an exciting starting point for the realization of this concern.

Predictive maintenance of the human body could be realized in this way. This technology is already being used for machines. By using sensors, it is possible to detect in an aircraft engine, when it is still working without dysfunctions, that a certain part is likely to become dysfunctional in the next six months. To prevent the danger, the part in question is replaced immediately. An analogous procedure can be realized with the constant monitoring of the human body by means of RFID chips. The prolongation of the health span can probably be significantly promoted in this way, since dysfunctionality can already be remedied when it is not yet identifiable, but only indications based on sensory measurements are present. At the same time, however, personalized health data on each person would have to be stored in a comprehensive manner, which poses numerous risks, depending on who has the right to access the respective datasets. After all, the data could be used against you, whether in the form of legal, institutional, or even moral sanctions. At the same time, however, comprehensive monitoring also goes hand in hand with the possibility of minimizing personal and social risks. The following examples illustrate this central insight explicitly.

In this way, an HIV infection could be detected promptly, so that appropriate medication could be prescribed immediately to minimize the physiological viral load. At the same time, this technology is of relevance for the containment of epidemics that can hardly be underestimated. Those infected with Sars-CoV-2 could be immediately detected, isolated, and treated if human beings upgraded by chips became the new norm.

We are not that far off. The widely used smartphones would have to be integrated into our bodies, instead of our just carrying them around as external devices. But even in the latter way, comprehensive digital surveillance can already be used efficiently in the event of an epidemic. China is already doing this in the case of the coronavirus pandemic. There, the implementation of Smart Cities is already being further advanced. Citizens receive a traffic light on their smartphones that is supposed to indicate the probability of an infection. If the light is yellow, citizens are obliged to isolate themselves for 7 days; if it is red, the isolation should last 14 days.[25] The assessment is based on the evaluation of big data. Which infected person has stayed where, when, and for how long? Who was in the affected areas? Who interacted with whom, when, for how long? If one also takes into account the behavioural patterns of the digital data of those actually infected, the probability of an infection can be calculated solely on the basis of a person's pattern of interaction. The associated possibilities to contain an epidemic are numerous.

We are constantly changing hybrid cyborgs. The latest technological innovations do not represent a breach in the dam as regards what we are allowed to do with our bodies. The latest upgrades are part of a long tradition that begins with language upgrades. A cyborg is a steered or upgraded organism, as the ancient Greek origin of the term indicates. Κυβερνήτης means helmsman. In 2016, researchers from Canada and Israel convincingly demonstrated that our bodies contain numerous non-human cells, that is, bacteria and unicellular microbes (Sender/Fuchs/Milo 2016). They are even said to make up the majority of cells in our body. In fact, human bodies are said to consist of more non-human cells than human cells. According to Ron Milo and Ron Sender of the Weizmann Institute in Rehovot, Israel and their colleague Shai Fuchs of the Hospital for Sick Children in Toronto, for every 30 trillion human cells, that is, cells with our own DNA, there are 39 trillion non-human cells. So we are hybrids, crossbreeds (Sorgner 2020d).

All these developments raise further moral challenges. The foregoing suggestions concerning how to revise the meaning of a person and what it means to harm a person are initial suggestions for a revised concept which takes into consideration the various events within the posthuman paradigm shift. It is not meant to be a final answer, but a suggestion for further reflection and debate. Inventions, scientific insights, and technologies radically alter our lifeworld. We need to take them into consideration and carefully reflect upon these changes.

It is true that with every innovation there is an increase in correlated risks. But so far we have managed to minimize risks and increase the likelihood of living a good life. The latest technologies also offer enormous potential for this, which must be used in a politically appropriate manner. Smart Cities need upgraded people. I can hardly wait for our posthuman future to occur.

4.4 Transhumanism and 'The Land of Cockaygne'[26]

A central element of most successful utopias is that they provide answers to some of the hardest human challenges. The poem 'The Land of Cockaygne', which was composed around 1330, represents a widely shared utopian dreamland. The bioconservative bioethicist Michael Hauskeller claims that 'Reinventing Cockaygne' (2012) is a central transhumanist goal. That utopia is of enormous importance, for transhumanism is confirmed by Nick Bostrom and his 'Letter from Utopia' (2008). I have some serious reservations concerning both of their positions, which I will explain here.

There are religious, social, philosophical, and many other types of utopias. A utopia can be a good place – an *eu topos* – or it can also be an *ouk topos* – a non-place, something which is not desirable, in which case it can function as a warning. Even though utopias usually serve as a method of exploring alternative futures, some utopias lie in the past. Novalis regarded the Middle Ages as a utopian ideal, Winckelmann ancient times, and Rousseau a pre-cultural natural world. Of course, even in their cases, past ideals are used in order to present new options for future developments. These aforementioned utopias are interesting from an intellectual perspective, but they were not effective as social movements or as worldviews which large groups have regarded as appealing.

More effective as social movements are utopias which present solutions for social challenges, like Marx's utopia, or others which are concerned with the ultimate destiny of humanity, as well as that of individual human beings, like many Christian suggestions. In particular the latter have been extremely successful, because they provide answers to the most difficult human challenge, namely to find a way of dealing with death, dying, and human finality.

These issues also made me turn towards philosophy from a very young age onwards, when I became a teenager, and they are still central to many of my philosophical endeavours. And yes, embracing, presenting, and advertising a weak Nietzschean transhumanism has to do with what I see as a plausible way of dealing with death, dying, and human finality. However, this does not have to imply that I am also embracing a utopia. There are a number of transhumanist approaches with strong utopian traces, but there are also others which can be classified as anti-utopian versions of transhumanism. My approach clearly belongs to the second category, even though one might be able to argue that anti-utopias are merely a specific kind of utopia. There are uses of the concept of an anti-utopia which simply stand for a utopia with extremely negative outcomes. This is not the meaning which I employ. In addition, what I refer to when I deal with anti-utopias is not simply a non-utopian approach. Anti-utopian approaches include non-utopian versions of transhumanism, but they have additional implications in so far as they are also strongly directed against utopias, as they represent an enormous danger for human flourishing. I particularly wish to stress what has beautifully been pointed out by the demons in the final scene of the music drama *On the Noise of the World*, *Vom Lärm der Welt*, by the most fascinating German composer alive, Sven Helbig, namely that we are doomed if we actually follow a utopia. I rather hold that the rejection of any kind of ultimate utopia promotes human flourishing. It is central to realize that living a good and a meaningful life does not have to include that your vision is commonly or even widely shared, which, however, is an essential element of utopias. It does not even have to be the case that what you live by, what you hold on to, and what provides your life with meaning is shared by any other being. To have a meaning in life, if this is what you need, does not have to imply that you subscribe to a specific religion, either. Yet it is this field of discourse which still contains many problematic arguments and polemics.

Germany's most influential living philosopher, Jürgen Habermas, is a prime example in this respect. He has presented several false reasons for discrediting transhumanism as a serious philosophical approach. His way of dealing with transhumanism reveals a widespread tendency, namely to identify transhumanism with a quasi-religious approach (Habermas 2014). This is what he was doing in hinting at a relationship between transhumanist reflections and that of a sect. However, there have been other thinkers who have stressed religious elements in transhumanist positions, and this has been done for a great variety of motives, such as the following.

1. One reason is to stress that, concerning their validity, transhumanism is in no way superior to the views presented by traditional religious communities. Both are supposed to rest solely on faith. This, however, is not the case for transhumanism.
2. Others use this association in order to justify their own Mormon transhumanist beliefs. I regard these positions as internally problematic.
3. Further thinkers use this line of thought to discredit transhumanism as a non-immanent position, for example many critical posthumanists stress that mind uploading implies a dualist anthropology which is similar to that of many monotheistic religions. This line of thought is not the case, as it is possible to affirm a naturalist anthropology and a functional theory of mind so that mind uploading can conceptually be thought within a naturalist framework. (More 2013, 7)

In any case, I do not think that reflections which identify transhumanism with a religion or a quasi-religion have any plausibility. Furthermore, the ideas presented in these arguments do not go along well with most basic premises of transhumanism, nor do they correspond to my way of thinking.

However, one does not have to subscribe to a religious outlook to present a utopian philosophy, and there are a number of transhumanists who argue in favour of utopias or whose thinking contains many utopian traces. Three examples in this context are: Nikolai Fyodorov's concept of the perfection of the human race, Robert Ettinger's 'Cryonics Institute', and 'The Abolitionist Project' by David Pearce. There is a lot to be said about each one of these projects. However, for pragmatic reasons, I will merely focus on a specific topic which is employed fairly often by a great variety of transhumanists, namely that of 'immortality'. It is also at the centre of many controversies about transhumanism from my point of view.

4.4.1 Immortality

With immortality we have reached a topic which has been of great importance in the history of philosophy. It is also one where misunderstandings arise very easily. Russell Blackford stresses that I 'am too quick (and too keen) to absolve transhumanism of any commitment to pursue physical immortality' (Blackford 2017, 203). He might be right, but I do not think that this is the case, because it seems to me that many transhumanists do not seriously consider the meaning of the word 'immortality', which is the reason for their using it.

It is a grave misunderstanding to claim that any serious transhumanist affirms immortality in any literal sense, as it is not the case that serious transhumanists strive for immortality, if immortality is taken in the literal meaning of the word. Given a widespread acceptance of several versions of a naturalist worldview among transhumanists, immortality cannot even be thought as a realistic option. Let us say, it will be possible to download your personality first onto a hard drive and then again into a new body: does this mean that you can be immortal? Of course, this is not the case. Even in this instance, you are living in a solar system which will exist for just another five billion years. Maybe, we will have been able to move to another solar system by then, so that human, or trans- or even posthumans can continue to exist there. Even in that event, we will not achieve immortality in this way, as it is highly likely that the movements of the universe will either come to a complete standstill or that the entire universe will collapse eventually and a black hole of infinite density will come into existence. How would it be possible for any human or trans- or posthuman to survive such a situation? Hence, there are plenty of reasons for claiming that immortality in the literal meaning of the word is not a realistic option.

Still, there are transhumanists who use the word immortality and refer to it as a transhumanist goal. How are such utterances to be understood? Immortality in these cases needs to be grasped as a specific type of utopia, not as one which can actually be achieved. Personally, I think that in most historical cases utopias were used not as realistic goals but in order to highlight the relevance of specific qualities, situations, or characteristics. Such utopias, if understood correctly, are unproblematic. However, it is up to further investigations to clarify whether this is how most people understand the notion of a utopia.

In the case of the word 'immortality', it hints at the relevance of a long and healthy life. It must also be noted that in most cases human beings do not aim for a prolonging of their life span but a prolonging of their health span, that is, of the duration of time during which they live a healthy life. By using the word immortality, the likelihood increases of getting media attention, of getting financial support, and of being talked about in many diverse social circles and circumstances, even in theology departments. Any transhumanist who wishes to be taken seriously employs the word 'immortality' for rhetorical reasons only. What is important in this context is not that immortality is a realistic goal, but that a prolonging of the health span is a realistic goal, which is affirmed by most human beings around the world, and, hence, deserves further attention. When discussing immortality and transhumanism, it is this issue which needs to be stressed. We talk about immortality in order to get public attention and to get funding for an event, but not because we regard immortality as a realistic option. Furthermore, it needs to be highlighted that affirming the importance of a prolonged health span does not imply a strong claim concerning a valid concept of the good life. It simply describes what most, but not all, people value. To affirm a strong concept of the good life would have to imply claims concerning what is necessary for all people for living a good life, which is not what I am doing, and which is a highly implausible goal.

4.4.2 An anti-utopian transhumanism

Instead of embracing and presenting a utopian transhumanism, I am arguing for an anti-utopian transhumanism. By means of utopian projects the danger exists that the present is sacrificed for a utopian future which will never happen and which might and most probably ought to be seen as a vision which cannot even be properly conceptualized. The German 'Third Reich', the various communist utopias, and mediaeval Christian political system clearly reveal the harm which can come about as a consequence of a utopian political project. Utopias are fine, if they are meant as rhetorical devices to hint at certain social and individual challenges. If they are meant literally and, worse still, if they are understood literally, then utopias can have problematic implications. Fundamentalists can use them to try to install

all types of totalitarian means to bring about the desired utopias, which, however, cannot be realized. In these cases, people and the present in general are sacrificed for an impossible future. Unfortunately, there have been too many such social experiments in the history of humanity already, and which have had dramatic consequences. Instead, my anti-utopian transhumanism focuses on and stresses the importance of realistic goals.

4.4.2.1 Dynamic politics of freedom, equality, and solidarity

My anti-utopian transhumanism stresses the central importance of the norm of negative freedom, not as a fact but as a social achievement for which generations of different interest groups have been fighting within the Enlightenment. It enables human beings to live in accordance with all of their idiosyncratic concepts of a good life. I am not claiming that it has been realized in the appropriate manner in any society, but it has gained more social relevance than ever before in many enlightened countries, at least. However, its relevance must never be forgotten and must permanently be fought for, if it is to be considered in an appropriate manner. However, this does not imply that equality and solidarity should not be important. This is definitely not a claim of mine. Still, I see the other two norms mentioned as derivative of the norm of freedom and in a dynamic dialectical relationship with freedom (see Sorgner 2015a, 225–236).

4.4.2.2 The importance of a radical plurality of goodness

The focus on the norm of freedom also has to do with my concept of the good. In contrast to other transhumanists who uphold a Renaissance ideal of the good or a common-sense ideal of the good, I regard any non-formal concept of the good as highly implausible and argue in favour of the radical plurality of goodness. Someone lives a good life by following their very own idiosyncratic psychophysiological demands, their very own desires, passions and fantasies. Most of us might not even be fully aware of what we want, because we are too strongly influenced by the one, Heidegger's 'man'. What one regards as appropriate gets taken over as a second nature by many people. One should not do this and that, and by living in any type of culture certain oughts or ought nots become the basis of a second nature, whereby we do not even realize anymore, in most instances, that these established habits do not correspond to what we actually desire. To become aware of one's very own drive is much more difficult than is often believed (Sorgner 2016b, 141–157). Yet, to embrace this concept of the good has significant cultural, political, and social implications. It concerns cases like the following: person A wishes to have her healthy leg removed, as she does not regard it as belonging to her. A radical concept of the good does not exclude the possibility that her wish can be an authentic one.

4.4.2.3 Affirmation of a culture of plurality, science, and relationality

If we stick to and promote the norm of freedom, we can, hopefully, even promote the coming about of a culture of plurality, and it is this direction which I also regard as appropriate for Europe, without claiming that there will ever be a perfect culture of plurality, from which time onwards we will all live blissfully ever after. Most traditional utopias, which are taken as realistic goals, go against this suggestion, as they demand more uniformity within a culture. In addition to this, we need to attribute a special role to scientific insights, but not because they provide us with the truth in correspondence to the world, but because they work and have proven themselves to work, as they are based on empirical studies. If insights are no longer plausible, due to further studies, more up-to-date research, and new evidence, then we need to give them up, develop further, and embrace new insights. It is this affirmation of flux, chance, and dynamic processes which is lacking in most traditional utopias which present an unchanging perfect future state to which we should aspire. Finally, it is the element of relationality which I am stressing and along with which goes the need to revise many ethical paradigms. This one is quite a complex thought, and therefore I will merely refer to one example in order to reveal the impact of this premise. It has to do with the implausibility of the culturally dominant way of conceptualizing humans, embryos, and other animals. Human beings count as subjects. Animals, in most legal constitutions, fall under the object law and are to be treated like things, with the German basic law being a paradigm case for this outlook. However, this understanding is no longer shared widely today, which leads to the need to revise the way we deal with these issues (Sorgner 2010a, 212–266). Herein, many transhumanists agree and stress the need to introduce personhood for non-human animals. Maybe this demand even needs to be expanded to personhood for non-human entities with sentience, and self-consciousness, and maybe even to post-personhood for entities with hyperautonomy, superintelligence, nano-sentience, and meta-self-consciousness. Consequently, there might be a need to attribute post-personhood to Data from *Star Trek*, if he came into existence. This example merely reveals the paradigm-shifting impact of this way of thinking.

It might also be argued that this anti-utopian thinking is regarded as a utopia, even though I rather stress that it is a pragmatic way of moving forwards without a final goal, by permanently looking for problematic circumstances which need to be fixed and then dealing with them. Yet, there are arguments to regard such an outlook as utopian, too; for example if one holds that humans have a widely shared need for utopias, then, paradoxically, this suggestion would have to count as a utopia, too. However, traditional utopian thinking, if it implies that the utopian goals count as realistic ones, usually has problematic paternalistic and totalitarian implications, as the ends can be used to justify any means, and, unfortunately, in human history we have already had too many examples of such political disasters. However, these risks are not given on the basis of the approach I am advertising (Sorgner 2017b, 193–200). One reason why many people today still desire new utopias is that they do not realize what wonderful achievements we have already realized. Many of these achievements are related to our technological inventions. Of course, for fully dealing with the realization of the following list of human achievements further explanations need to be considered too, for example colonialism, imperialism. Let me just briefly refer to three particularly striking examples that highlight the widespread ignorance of such achievements.

4.4.2.3.1 A decent work–life–balance with a lot of vacation time

In Europe we are decelerated, as we have never been before. However, a thinker such as Hartmut Rosa stresses the need to slow down even further (Rosa 2014); the permanent demand to decelerate is his mantra to all social challenges. If you see and realize how people work in Silicon Valley and in East Asian countries, then it is hard to grasp how anyone can take this suggestion seriously. In contrast, I regard it as important to realize that we have never had as much free and vacation time as people in Europe have today. There have never been more people who do not have to struggle for the basic necessities of survival as there are in Europe today, too. Social welfare, universal public health insurance, and good education systems are of central relevance for this being the case. It is an utterly significant insight to realize that this is the case. Decelerationists have the utopian fantasy that we all could dedicate ourselves to doing pottery in Tuscany all the time, but this is not how life works. Ivory-tower scholarship is not helpful in circumstances when the need arises to take the perspective of a globalized world into consideration, which reveals that, if we decide to slow down further then the most probable future which Europe will have will be that it will become the world's Disneyland for Americans and Asians. Maybe this is the case already.

4.4.2.3.2 Non-violence as a social and a lived ideal

Sometimes mass media seems to tell us that everything is going downhill. There are wars and bad crimes happening everywhere and all the time. However, it is important to realize what scientists have to say about this claim. Here we have to note that the level of physical violence has never been so low as it is today. Of course, the situation is not a perfect one, and there could always be less violence. However, if you carefully consider the ground-breaking and convincing research undertaken by Harvard psychologist Steven Pinker, then it will become clear that human rights, liberalization, and democratization have led to a world with less violence, child abuse, domestic violence, and other types of cruelty than ever before (Pinker 2011). It shows that the situation is not as bad as many think it is, in particular here in Europe. Furthermore, it might be worth remembering that it is plausible to claim that there is a correlation between cognitive capacities and the likelihood of acting morally. By developing further in regard to our cognitive capacities, the likelihood increases that the level of violence in our society will be reduced even further.

4.4.2.3.3 An increased lifespan or even better: a longer-lasting health span

Last but not least, it is the health issue which needs to be addressed. Here, we return to the transhumanist utopia of immortality. When talking about the use of the word 'immortality' as a rhetorical device, I have hinted at a type of utopia to which my utopia-criticism does not apply, as here the idea is merely being used as a way of advertising a certain insight, and not as an idea which is supposed to function as a realistic goal.

A prolonged health span is a realistic option and various enhancement technologies provide us with several diverse options for realizing this goal. That our life expectancy is not fixed becomes clear when considering the various life expectancies in countries worldwide today, which vary from less than 40 years in some countries in the sub-Saharan parts of Africa to more than 80 years in many countries in Europe, North America and Australia. Furthermore, the average global life expectancy has also significantly changed over time, from an average of 50 years in 1960 to an average of 65 years in 2010. Historical research also confirms important changes. It shows that the average life expectancy in ancient times was even lower than this. Scientific research confirms that the increase of our life expectancy is related to a great multiplicity of enhancement technologies, from hygiene, via education, and vaccinations, to contemporary biotechnologies. Consequently, there are reasons for claiming that technologies have been extremely successful in promoting the human life span, and also our health span, which again provides us with a reason for expecting that emerging technologies will continue to help us in this respect, so that this development will continue. By focusing our scientific research on the right questions, it can be expected that this development will continue in an exponential manner. Here the central question is the following: what are the best possible research topics in this respect? Aubrey de Grey (2007) stresses the relevance of the seven deadly sins of ageing. Ray Kurzweil refers to the option of mind uploading (2006). Others might stress the potential of artificial superintelligence, cryonics, genetic modification, or genetic selection procedures after IVF and PGD. It is highly likely that no single answer will be the right one, but that a great variety of technologies will be able to influence and prolong our life expectancy.

Personally, I regard the field of genetics as most promising in order to prolong our health span. It might be interesting to note that even a bioconservative like Habermas, in the context of discussion on gene ethics, acknowledges the relevance of longevity. In some passages of his essay on liberal eugenics he talks about preventive measures as well as a prolonged life expectancy in the context of morally legitimate therapeutic purposes of eugenic intervention, as all these measures can be seen as all-purpose goods (Habermas 2001, 48). Unfortunately, he is not consistent within his reflections, because a couple of pages later he claims that genetic interventions can be justified only to avoid extreme evils or illnesses. By solely focusing on his former remarks, Habermas could be seen as a transhumanist thinker, because enhancement measures can be interpreted as specific cases of therapy. However, such a reading is clearly not in tune with what Habermas has in mind. Still, it needs to be noted that even for a bioconservative thinker such as Habermas, it is not out of the question to employ gene technologies in order to prolong the human life span, which reveals the widespread importance of this goal and the praiseworthiness of transhumanists who bring this goal into the right focus by employing special rhetorical methods, for example the use of words such as immortality. It is also a timely endeavour, as scientific research reveals that the presence of certain genes in specific circumstances seems to indicate a higher likelihood of longevity, for example CETP (cholesteryl ester transfer protein) genes. What is the relevance of such findings?

If longevity or a prolonged health span is identified with a widespread or even an all-purpose good, then there are reasons for allowing genetic enhancements in order to promote these genes. There are several options for how this can be done: (1) genetic enhancement by modification: studies reveal that such modification can be done without there having to be any side-effects, and I have shown that, morally, such modifications ought to be evaluated in the same way as parental education (Sorgner 2015b, 31–48); (2) genetic enhancement by selection after IVF and PGD: again, it is clear that this procedure is already a reliable one, but also one which is still forbidden or whose use is radically restricted in many countries, for example Germany (Sorgner 2014b, 199–212); (3) pharmacological, cyborg, or morphological enhancements which take into consideration a gene analysis: it is this option which can gain wider acceptance due to the enormous amount of information which is being collected, due to big gene data. The impact of this research and the relevance of related moral issues such as that of bioprivacy and gene privacy can hardly be overestimated (Sorgner 2017b, 87–103). However, these reflections are meant solely to hint at some small area in which enormous developments have already taken place. Many other examples could be mentioned, too.

4.4.3 Conclusion

There are two types of transhumanists. Those, who affirm a utopian perspective as a realistic goal, and others who do not. The fantasy, dream, or, for some, nightmare that there will ever be a world without suffering, full of high jinks and without any fear of not even being able to survive on an everyday basis, is one which I regard as highly dangerous, because it has been abused too often to justify limitations, violent actions, and paternalistic attitudes toward currently living human beings. As a counter suggestion, I have sketched the outline of an anti-utopian version of transhumanism, which functions on the basis of some basic guidelines. These could help us to detect morally problematic current structures so that we can try to get rid of them. The longing for a utopia often arises due to discontent with the present, which again is very often associated with overly high expectations for one's own life. Permanently doing pottery in Tuscany, or a life in the Land of Cockaygne, are not realistic options. By focusing on the manifold achievements which we have already been able to realize, it might be easier not to be tempted into following a life of silly daydreams. I am not a disillusioned romantic or cynical depressive. However, I think that by embracing an anti-utopian stance we can achieve more, lead more fulfilled lives, and decrease the likelihood of being doomed. I think there is a lot to be said concerning all of these goals.

4.5 Transhumanism, immortality, and the meaning of life

Asking for a meaning in life, we seem to seek for something which is relevant for us, but which transcends our own limited life span. God can provide us with meaning, as God has created the world and us, and God is the truth, and can enable us to have a blissful afterlife. Yet, we have no clear indication whether God exists or not.

C.G. Jung stresses: 'Man cannot stand a meaningless life' (Jung in Stevens 1994, 126). Yet, he realized that many human beings past their mid-life crisis, who have families and flourishing careers, are tormented by the meaninglessness of their lives. Everything is bound to dissolve, and fade into nothingness. Nothing will remain from us after we will have died, which will necessarily happen to all of us. Our offspring, the books we have written, as well as the artworks we have created, are all bound to dissolve eventually. We spend all of our lives working, struggling to survive, changing the world, running an institution, being politically active … until a virus comes and wipes all of us out, or a war, a car accident, or a fatal disease. Many humans see their offspring as a continuation of their own lives, but the same destiny applies for them, as well as to their offspring. Even if your pedigree were to survive for another five billion years, then it will be over; just as, with the ending of our sun, our solar system cannot exist either. Maybe we will have escaped from this solar system by then. Yet no solar system will exist forever. What about the entire universe? The expansion process of the universe might come to a standstill and any movement might fade away. Maybe, the expansion process of the universe will be reversed, such that the universe will collapse and a cosmological singularity will occur, a point of infinite density. None of our offspring will survive such a situation. Seeking for a meaning in life by having offspring does not seem a plausible solution to what we are looking for, neither is the option of personal immortality in this world a plausible one, as all of our lives are bound to end eventually. Even if I managed to upload my personality to a hard drive and then download it again into another carbon-based body, maybe even into a body of a different shape or species, this would not provide me with immortality.

Immortality can have two different meanings. An immortal entity either cannot die or does not have to die. Being alive in this world, it is hard to imagine how there could be an entity which cannot die. If you do not have to die, but can die, given the appropriate circumstances, these circumstances will occur eventually. We need to be aware that the chances of infinite survival are practically non-existent. What, then, is the relevance of mind uploading and cryonics, which are being affirmed by many transhumanists, given these circumstances? It is this question with which I will be concerned in the first and second parts of this section. Both of these technologies increase the likelihood of our having an increased health span, which most people identify with a higher quality of life. Even Nietzsche, whose dynamic thinking is structurally analogous to that of many transhumanists (Sorgner 2009), implicitly points out that the majority of people identify a flourishing life with longevity. Nietzsche acknowledges that many people want to increase their health span, as this is how he characterizes 'the last man' in *Thus Spoke Zarathustra*. However, he affirms a different lifestyle, one which is more experimental, and one which aims for the special moment by means of which one can justify one's entire life, rather than maximizing one's health span. This rationale has to do with his theory of meaning, the eternal recurrence of everything, which seems to be the most plausible alternative theory of a meaning of life to the religious afterlife theories offered by monotheistic religions. In the third part of this section, I analyse the search for an alternative theory of meaning. On the basis of non-dualist transhumanist reflections, no meaning can necessarily be established, which leads us to the fourth part, in which the relationship between perspectivism and the question of a meaningful life can be clarified further.

4.5.1 Cryonics

Even the most optimistic transhumanists regard the possibility of mind uploading as being a few decades away, be it Kurzweil (Kurzweil 2006) or Musk.[27] However, each one of us continually has to face the possibility of dying. What if the option of mind uploading, which could help me to increase my life span, becomes available too late? There so many chances for death? In a world without a hope for an everlasting afterlife, death is final. Never seeing one's friends and loved ones again. Never experiencing the kaleidoscope of pleasures again. Never feeling the exciting power of shaping the world again. Something needs to be done. Here, cryonics becomes relevant. Cryonics provides us with a hope, even if there is no other hope, given the current state of medicine (Doyle 2018). Cryonics offers a minute chance of coming back to life, even though you must have been clinically dead for cryonics to be an applicable option for you. Cryonics does not demand the existence of an immaterial world, a world outside of time and space, but it is a technical option which could work in principle.[28] There have been successful trials on some animals.[29] Still, the idea sounds more like science fiction than like proper science, as it seems to involve the possibility of resurrecting someone after that person was clinically dead. This is the crucial issue. We need to consider and reflect upon what it means to be dead.

Whether you are dead or not depends upon the currently valid definition of death, and the definition of death has changed in the past 50 years (Youngner 2002)[30]. It used to be the case that the heart had to stop beating and, if within a couple of minutes it was not possible to detect that it spontaneously restarted, then a person was declared dead, which is associated with the irreversible cessation of circulation and respiration. This is still the most widely used criterion of death. The heart is no longer beating, there is no blood pressure, and there is no blood circulation. You encounter rigor mortis, and about 20 minutes later decomposition initializes. This is a clear sign of someone being dead, and it used to count as the decisive criterion for someone to be declared as dead.

However, as a consequence of technological innovations, the definition of death had to be altered. An additional death criterion has been added to enable the possibility of harvesting organs for transplantation purposes. The criterion of brain death was introduced and widely accepted. If there is an irreversible loss of all brain functions, it is impossible for a person ever to regain self-consciousness. This criterion of death has been adapted, too.

Yet, it needs to be kept in mind that the criterion of brain death does not apply in everyday situations. In most cases of death, the aforementioned criterion is still the decisive one. The criterion of brain death is considered only in cases where intensive care is being applied and someone's circulation and respiration are being sustained by technological means. Even if the brain were dead, intensive care would keep the respiration and circulation going. Hence, different methods for deciding upon someone's death are needed. They are considered if there are reasons for claiming that the brain has been significantly damaged. Certain reflexes can be tested, for example the corneal reflex. If the eyelid does not close, it can be a sign that brain death has occurred. Yet, this is not guaranteed. Additional tests are more reliable, for example if there is no blood in the brain, and you can check this by using ultrasound, computed tomography, or an electroencephalogram, which are all reliable ways of testing whether brain death has occurred.

Full brain death is the decisive criterion for regarding someone as clinically dead, and only if this is the case, may the corpse be buried or cryopreserved. Only then may the process of cryonics be initialized. The blood needs to be removed, the veins cleansed and then filled again with another liquid, which enables the body to get vitrified after having been placed in liquid nitrogen of –196 degrees Celsius (Doyle 2018, ch 6). This can be done with the entire body or merely with the head. The hope associated with the process is that in the future the cessation of all brain functions will no longer be seen as an irreversible state. If this were the case then the body could be unfrozen, and the functionalities could be restored, such that the body would be alive again. An alternative option is that the personality is preserved by means of cryonics. In this case, it might not be necessary to revitalize the ceased brain functions, but one would have to transfer the personality to a hard drive, such that the person could continue to live as an uploaded mind (Doyle 2018, chap. 6, 20), which raises further challenges concerning the concept of mind uploading, which I will address next.

The main pragmatic challenges which this process has to deal with are that decomposition of the body starts very fast, and that further damage can easily occur during the preservation process. Both are significant issues. On the other hand, it needs to be noted that you do not have much to lose if you let yourself be registered for cryonics – merely some inheritance money might be lost. If you are dead and do not use cryonics, you do not have any chance of becoming alive again. If you use cryonics, there is a chance of becoming alive again, no matter how small that might be. If money is not an issue, then it might be worth taking up the option. Still, the practical challenges are significant. What can you gain? You could either become alive again in your old body or continue living as an uploaded mind. Still, I regard it as important to keep the following in mind: you will not gain immortality as a consequence of cryonics, and this could not be achieved by means of mind uploading, either. So, what can be said about the possibility of mind uploading?

4.5.2 Mind uploading

Can we transfer our carbon-based personality onto a hard drive so that nothing is lost? A digital entity so far relies on a silicon basis, which is structured such that it consists of ones and zeros. If our personality, which is connected to a carbon basis, is structured non-dualistically, the process of transformation to an uploaded mind cannot occur without any loss, as digital entities are structured as ones and zeros. This worry is a fundamental one. So far, it is unclear which option is is better founded. Not even the natural sciences provide us with solid reasons for regarding one option regarding the structure of energy as more plausible. On the one hand, energy can occur only as an integer multiple of the Planck constant. If this is a decisive insight, then the entire world could be organized digitally. All entities could then be analysed as ones and zeros. However, the wave-particle dualism of quantum physics reveals that other interpretative options are also plausible. The technological possibility of mind uploading cannot in principle be ruled out for this reason alone (Kind 2020, ch 6).

Do we have a reason for claiming that a living uploaded mind is in principle possible? This is a further challenging issue, as we are confronted with the philosophical question of what counts as being alive. All living entities which we know so far consist of cells, are self-moving, and consume organic matter, that is, they possess a metabolism. Are there any digital silicon-based entities which have these criteria? A self-replicating digital entity which comes to mind is a computer virus. If the capacity of self-movement were sufficient for being alive, then we could indeed regard computer viruses as being alive. Stephen Hawking held this position.[31] However, a computer virus does not consist of cells, and does not possess a metabolism. A non-digital virus does not count as being alive, for the same reasons. A virus merely replicates within a cell of a living organism.

So far, I have considered only the case of the computer virus, as hereby we might have the strongest example of being confronted with a digital entity which could count as being alive. Upon further reflection, this does not seem to be the case, as it does not possess many central qualities of living entities. Why, then, should it be at all plausible that mind uploading is a likely process, which, according to some, will occur in a few decades? So far, I have considered only the possibility of there being a living digital entity. Yet there is no digital entity which fulfils all the central criteria of being alive. However, being alive is not even a sufficient condition for mind uploading such that a personality transfer could occur without any loss. We would want to continue being competent, self-conscious, living entities after the upload. Yet, not even the minimum criterion for being a living digital entity has been realized so far.

A counterargument to this line of thought might stress that these reflections do not sufficiently consider the intelligence explosion, which goes along with the exponential growth in digital technologies. This judgement can be supported further by reference to Moore's law. This might be a sound argument, and I cannot exclude the possibility of uploaded minds. However, given the aforementioned reflections, it seems highly implausible that any such development will occur in the near future such that mind uploading could help to increase our life spans, which is the main reason why most humans are interested in this technology.

Other technologies, like gene editing or cyborgization, are far more promising for promoting the likelihood of an increased health span (Sorgner 2018b, Sorgner 2019c). Without underestimating the power of digital technologies, the widely shared talk about uploaded minds reminds me of the mediaeval discussion on how many angels can dance on the head of a pin. Both mind uploading and the size of angels are topics which make sense on the basis of the currently dominant paradigms. However, neither of these issues can be regarded as being among the most pressing issues of our time. Affirming mind uploading might help you to get onto the cover of a magazine, but it might also distract the intelligentsia from the really pressing issues which they ought to be concerned with.

Both cryonics as well as mind uploading are technologies which could enable us to radically increase our life spans. This is what needs to be kept in mind. Most humans equate an increased life or, more precisely, health span with a better quality of life. If a technology can increase the likelihood of increasing the life span, then it is technology which also, politically, ought to be taken seriously, as it could promote the good life of most people. However, it does not provide them with an answer to the question of the meaning of life. I will reflect further on this issue in the next section.

4.5.3 Meaning of life

Cryonics and mind uploading are widely associated with transhumanism. Yet neither of them can lead to immortality, a concept which is often mentioned when these technologies are being discussed. Thus, the possibility of providing a transhumanist meaning of life cannot be achieved.

Nietzsche was a philosopher who explicitly tackled the question of meaning, and he is one of the few philosophers whose reflections can be seen as structurally analogous to transhumanist thinking. The main reasons for this are his naturalist ontology of continual becoming, and his serious consideration of evolutionary thinking which made him affirm the concept of the overhuman (Sorgner 2010a). At the same time, he was struggling with the concept of meaning, having been strongly aware of the force and impact of the Christian concept of meaning which is associated with a personal afterlife. Whatever we do is relevant and has implication for the real life, the life after death, which does not merely last for a few decades. Here, we clearly have a meaning, whereby a meaning is something which lies beyond the human but which is relevant and important for humans without its being able to fade away like all other this-worldly endeavours.

The thought Nietzsche played around with was the eternal recurrence of everything (Sorgner 2016e, sec 16), which he clearly held as a cosmological theory. I am not saying that it is not relevant for ethical issues, but he affirmed it, as he regarded it as a plausible theory for scientifically minded thinkers. He even considered moving to Paris to study physics so as to prove the eternal recurrence scientifically himself. In his notes, not published by himself, he put forward several reflections and arguments by means of which he tried to present the plausibility of the eternal recurrence of everything. None of his lines of reasoning was conclusive. However, it needs to be stressed that it is not the case that no such line can be presented (Sorgner 2007, sec 102). The central thought which needs to be applied is the attempt to eliminate infinities when thinking about ontologies, as no infinity can be meaningfully conceptualized. No infinity can be completed. By doing so, the following judgements arise.

The total amount of energy is finite.
Energy can turn up only in a finite amount of states (properties). Modern physical theory which holds that energy can appear only in a quantity which is the integral multiple of the Planck constant supports this position (Energy can only appear in a quantity $= n \star$ Planck constant; $n = 1, 2, 3, \ldots$).
Time and space must not be independent of energy, but do not exist as separate infinite entities.
The law governing the energy must be a determined one. Scientists attempt to realize a great unified theory, which tries to find a theory by means of which the four fundamental forces can be unified: the strong, and the weak force, the electromagnetic as well as the gravitational force.
The reversibility of all energy states must be given. Einstein's theory of relativity seems to allow this.

If these premises are given, then it is necessary that at a certain moment all possible combinations of energy must have occurred. If one state recurs, the reversibility of all states is not excluded, and the law governing the energy is a determined one, then it is necessary that the cycle of the eternal recurrence has to begin anew, and all events, which have occurred before, need to occur again. However, at this point we run into a linguistic challenge, as, philosophically speaking, it is false to claim that events have occurred before. If events occurred before, then we do not take identity in the strongest Leibnizian sense seriously. If recurrence of the same implies an identity in the strongest Leibnizian sense, we do not end up having the nth circle plus the n+1st circle and the n+2nd circle, but we always have to have the same circle. If we talk about recurrence, we still have a linear concept of time in mind. However, if the aforementioned premises are plausible, we would end up with a circular concept of time, from a to z to a to z.

Which implications would this concept of time have for the question of the meaning of life? If we become aware of the eternal recurrence, we know that whatever we do will recur infinitely often. Everything which occurs is necessary, but, as we do not know what will happen, it seems to us as if it was possible for us to freely decide what we wish to do. Free will is merely the uncertainty of knowing what will happen next. Not making a decision or not acting is not an option. Not acting or acting are decisions you are making. Hence, we still try to reach our goals as we have always done, while thinking that whatever occurs is necessary, and it has to occur.

Furthermore, it is possible that the eternal recurrence can change your preferences. Most human beings identify a long and healthy life with a higher quality of life. However, living a long and healthy life without ever realizing what is important to you might not be a better life after all. Instead of focusing on expanding the health span, Nietzsche regarded it as central to experience one situation by means of which he could justify his entire life, and whatever had happened to him beforehand or what would happen afterwards, as the awareness of the eternal recurrence then leads to an altered state of awareness. You know that you have experienced that special moment, which is worth all the suffering you have to endure, as you know that this one special moment will recur again and again. *Amor fati.* Hence, experiencing such a moment provides you with a meaning in life. This thought by itself might enable a person to bear all the suffering each one of us has to go though. One moment can justify your entire life, and provide you with meaning. Nietzsche stressed that an expanded health span is what the 'last man', the majority of human beings, strive for. However, the ones who have grasped the challenging insight of the eternal recurrence will seek the moment by means of which they can give meaning to their entire lives. Spinoza held a comparable concept with his *amor intellectualis dei* (Yovel 1986). Goethe's Faust also sought a comparable goal when he made the contract with the devil.

> If ever I to the moment shall say:
> Beautiful moment, do not pass away!
> Then you may forge your chains to bind me,
> Then I will put my life behind me,
> Then let them hear my death-knell toll,
> Then from your labours you'll be free,
> The clock may stop, the clock-hands fall,
> And time come to an end for me!
>
> (Goethe 2008, lines 1698–1706)

Whether Mephistopheles fulfilled this demand or not is a tricky issue and depends upon the reading of Faust 2, in which Faust says the following: 'Zum Augenblicke dürft' ich sagen: Verweile doch, du bist so schön! Es kann die Spur von meinen Erdentagen Nicht in Äonen untergehn. – Im Vorgefühl von solchem hohen Glück Genieß' ich jetzt den höchsten Augenblick'.[32]

It also needs to be noted that the thought of the eternal recurrence has already been held by Heraclitus, the Pythagorean Eudemus of Rhodes, and some Stoics. Here, you see what Heraclitus claims: 'There is a [Great Year] whose winter is a [great] flood and whose summer is a world conflagration. In these alternating periods the world is now going up in flames, now turning to water. This cycle to consists of 10,800 [years]' (Kahn 2008, fragment XLIIIA, 49, see also: Raju 2003, 503). Eudemus of Rhodes explicitly states the following: '"Everything will eventually return in the self-same numerical order, and I shall converse with you staff in hand, and you will sit as you are sitting now, and so it will be in everything else, and it is reasonable to assume that time too will be the same", H. Diels and W. Kranz, Fragmente der Vorsokratiker, 58B34' (Capek 1991, 276).

Various Stoics are associated with the idea, too: 'Zeno, Cleanthes and Chrysippus hold that matter undergoes transmutation, as for example fire turns into seed, from which once again is restored the same world order, that was before' (Sambursky 1976, 168). Chrysippus, in particular, is mentioned in this context:

> Chrysippus, in his first book On Nature, says: The transmutation of fire is as follows. It turns first into air, then into water; and from the water at the bottom of which earth settles, air rises. As the air becomes rarefied, the aether is spread around in a circle; and the stars and the sun are kindled from the sea. (Sambursky 1976, 168)

It is important to realize that the circular idea of time has been shared by quite a few significant thinkers in the history of philosophy, but it is mostly absent in contemporary discourses, which strongly hold on to a more intuitively plausible linear account of time. However, it might be the case that we merely affirm a linear conception of time due to our ways of thinking being strongly influenced by our Christian cultural past, which demands this way of thinking, as otherwise we could not meaningfully wait for the Parousia of Jesus Christ.

What we can infer from these reflections is that the concept of the meaning of life has significant implications for our concept of the good life. If there is no meaning, we might merely be like Camus's Sisyphus, permanently dealing with our painful efforts, which we can accept (Camus), affirm (Nietzsche), or reject (Schopenhauer). If our meaning lies in a blissful afterlife, we need to do our best to realize it, which usually means to fulfil God's demands and hope for God's grace. The eternal recurrence is the most plausible non other-worldly account of a meaning in life. If this applies, then Nietzsche's suggestion of a good life is at least a plausible option: to aim for a moment which is so glorious that we can justify all the great amount of suffering we have to go through in life. Could the aforementioned concepts be plausible for transhumanists, too? In the next section, I will address this issue further.

4.5.4 The hermeneutic circle of naturalism and philosophical interpretations

Most transhumanists are naturalists, according to an IEET survey. Being a naturalist can have different meanings. It usually implies that one regards empirically non-accessible entities as non-plausible, as one is concerned with whatever is empirically accessible. However, this does not apply to all transhumanists. Some affirm other types of ontologies, for example there are Mormon transhumanists. It seems clear that there are different types of ontologies which go along with being a transhumanist. Transhumanism implies neither naturalism, nor a different ontology, even though it has to be noted that most transhumanists are naturalists. However, all transhumanists affirm that it is in our interest to move away from the current human limitations through technological means, because it increases the likelihood of our living a good life. This breaking away from our current limitations can also be identified with the demand to increase the likelihood of the coming about of the posthuman. Please also keep in mind that there is not one specific meaning associated with the posthuman in transhumanist discourses. There are further meanings of the posthuman which come up within critical posthumanist discourses. Within transhumanist discourses, the posthuman can stand for a member of the human species who has significantly moved beyond the current boundaries of who we can be as human beings. The posthuman can also be a member of a new carbon-based species. The most radical concept of the posthuman is that of the uploaded mind, that is, a silicon-based entity, which either has a causal connection to a specific human being or which comes about as a consequence of AI evolution. Furthermore, it needs to be noted that it is not the case that you have to affirm mind uploading in order to be a transhumanist, as otherwise Julian Huxley, who coined the term transhumanism, could not count as a transhumanist (Huxley 1951).

Being a transhumanist means being positive concerning the use of technologies. Transhumanists hold that the use of technologies usually improves the quality of life. Being a transhumanist is not necessarily connected to a specific ontology, even though most transhumanists affirm a kind of naturalism. If this is the case, then it might be worth considering further which epistemology is most appropriate for a naturalist outlook, as a traditional correspondence theory of truth can no longer be plausibly upheld if a type of naturalism is being affirmed. According to the correspondence theory, it is the case that a judgement is true if it corresponds to the world. However, if the world consists of an ontology of continual becoming, and this is what naturalism normally seems to imply, then it is impossible that a correspondence theory of truth can be upheld, as it is impossible for a propositional statement, which consists of unchanging words, to correspond to an ontology of continual becoming. Something stable cannot correspond to something continually changing. In a solely naturalist world, there is no empirically inaccessible realm of pure being, or of Platonic ideas. Yet these would be needed for there to be a chance of grasping a propositional truth which corresponds to the world.

In an evolutionary world of continual becoming there would not even be motivation anymore for searching for the truth in correspondence with the world for the sake of it. We are a specific type of animal which longs to survive, strives for power, and desires pleasure. Truth for the sake of grasping the truth can merely count as a plausible human drive, if you assume that knowledge is something divine and we use our divine capacities for grasping the truth, that is, God's image in us becomes active by our reflecting upon the truth.

If this description of who we are is plausible, then perspectivism follows (Sorgner 2007). Every philosophical judgement is an interpretation. Being an interpretation implies that it can be false, but not that it has to be false. It is plausible that what we identify with knowledge is merely a tool which helps us to archive our idiosyncratic human goals, which are relevant for us. Yet, even this judgement is not an unchanging insight. Being an interpretation means that perspectivism also can be false but not that it has to be false. Falsifying a claim implies a truth criterion. Yet, if you lack such a criterion, then it also applies to perspectivism that it is an interpretation which can but which does not have to be false. However, it has to be noted that so far perspectivism has not been falsified, which again supports the plausibility of perspectivism. At the same time, it needs to be noted that if this is the case, we no longer have a necessary reason to reject monotheistic religions either.

It is possible that Catholicism is true. At least, it is a possible interpretation of the world. So far, many transhumanists, who have a naturalist inclination, do not believe this to be the case, as they rely only on empirically accessible information. As Catholicism demands an afterlife and an empirically non-accessible God, these are seen by naturalists as entities whose existence does not need to be posited, as there is no evidence for their existence. Still, it is not the case that their claim can be falsified. They can be rendered implausible, but not falsified.

The preceding narratives claim that the claim of the existence of a monotheistic God is implausible, as it demands the existence of empirically non-accessible entities, but not that it is false. In the same vein, it can be claimed that it is not false that we are already living in a computer simulation. Again, the same logic applies. That we already live as uploaded minds on a hard drive cannot be demonstrated to be false; however, we lack any good reason or evidence for this to be plausible. If there is no good reason in favour of making such claims, we should avoid them.

Hence, it can be consistent to believe in a monotheistic religion and to be a transhumanist. You might need to believe in even fewer empirically, not directly accessible entities if you are a transhumanist as well as a Buddhist, naturalist, or sceptic. Being a transhumanist merely goes along with a certain positive attitude, but does not necessarily imply any specific meaning of life. As most transhumanists are naturalists, it might, however, be advisable to consider alternatives to an immortal soul and a personal afterlife as a meaning of life, if a meaning of life is necessary for a life to be a good and flourishing one. The eternal recurrence is not the least implausible alternative for a meaning of life, if you consider non-dualistic alternatives to traditional concepts of the meaning of life, which are associated with personal afterlives.

Even though there are a great many discussions on immortality by transhumanists, such a concept cannot plausibly be upheld (Rothblatt 2015). These discussions show that a this-worldly concept of immortality is highly implausible, unless you consider eternal recurrences as a possible option. A personal immortality which implies that it is impossible for you to die in this world is not a realistic option, given what technologies can realistically realize today. A personal this-worldly immortality cannot even be meaningfully conceptualized and is not a philosophically meaningful thought.

This does not mean that using the word 'immortality' as an advertising tool is of no relevance. Making people aware that *vita brevis est*, and that it is politically relevant to promote our personal health spans, as this is what many people identify with a higher quality of life, is a worthwhile endeavour indeed. However, mind uploading or cryonics, even if they worked, cannot provide you with a meaning in life, they could merely increase the quality of someone's life. This is an important goal, but it does not provide you with a meaning of life, nor is it a religious goal.

Transhumanism is not a religion. You can be religious and be a transhumanist, even though most transhumanists are non-religious. Transhumanism is a philosophically informed positive attitude concerning the usage of technologies, whereby the positivity is based upon solid empirical studies as well as philosophical reflections. In *The Better Angels of Our Nature*, Steven Pinker (2011) has put together a wide range of developments which reveal how much better our lives are now in comparison with how humans used to live in the past. Our health spans have never been so long as they are today. However, there is still a lot of potential for increasing our health spans even further. This is what it means to possess a positive attitude, and this positive attitude has solid support from empirical and historical studies. Transhumanism is characterized by a positive attitude concerning technologies, but not by a specific ontology. Positivity concerning technologies is justified by reference to historical studies concerning the quality of lives which take into consideration how long people live, how high the absolute poverty rate is, and how widely shared democratic political systems are. In all of these respects, we are in a better situation now, than we have ever been in the past. These are sufficient reasons for embracing positivity concerning technologies.

4.5.5 Conclusion

Transhumanism is characterized by positivity concerning technologies, as transhumanists hold that technologies usually increase the likelihood of people living good lives. This also applies to cryonics or mind uploading. Upon further analysis, it has become clear that neither of these processes could provide humans with a meaning in life. If they worked, and neither of these technologies might be among the most promising ones on the basis of my analysis, they could improve the quality of a life, but not provide someone with a meaning in life. If Jung is correct, and humans need a meaning in life, then people have a wide range of options, as transhumanism is not necessarily identified with a specific ontology, even though empirical surveys reveal that the majority of transhumanists have a naturalist inclination. The eternal recurrence of everything, as affirmed by Nietzsche, Heraclitus, and some

Pythagoreans might be the most plausible non-dualistic meaning of life. If Jung is correct in his claim that humans cannot stand a meaningless life, it might be worthwhile for transhumanists to further consider this theory as a possible meaning of life, which could have significant consequences for their concept of the good life, too.

5

The End as a New Beginning

In these reflections, I have focused on philosophical issues concerning the insight that we have always been cyborgs. Technology has always been a part of us, and this applies to the emerging technologies, too. The judgements 'Nature = good', and 'Technology = evil' are part of a traditional dualistic mindset which is no longer plausible and which has many dangerous implications. This does not mean that there are no dangers connected to emerging technologies. The more efficient a technology is, the better the advantages, but also the greater the risks. These can be terrible, given inappropriate political circumstances.

I regard the implementation of a new type of paternalism as highly dangerous. A relational ethics has such paternalistic implications. With an awareness of the terrible paternalistic structures of the German 'Third Reich', I am convinced that everything must be done to avoid the occurrence of such frightening paternalistic political structures. Relational ethical approaches have such dangerous implications. I regard individual personal freedom as a wonderful achievement which must not be undermined.

There are critical posthumanist approaches which argue that it would be best if humans died out. There are other such approaches which demand that human existence on Earth must be regulated such that the relational complex of the Earth lives in an appropriately attuned order. This, however, demands that eugenic practices need to be implemented. We would need to forbid people to procreate other people. This undermines the wonderful achievement of negative freedom for which we have fought on various levels during the Enlightenment process. Scientists, intellectuals, as well as the wider public have fought for their rights to live in accordance with their idiosyncratic wishes, longings, and desires, and I regard plurality and negative freedom as wonderful achievements; I am happy that this insight is widely shared today. If you start from this insight, however, then it can be more problematic to deal with some global challenges like climate change.

Instead of the demand to introduce new eugenic laws concerning procreation or to get rid of human beings or to return to a natural world before the time during which evil technologies destroyed our harmonious relationship with nature, we desperately need to focus on technological solutions for the various issues which can be associated with climate change, for example in vitro meat; roofs made of solar panels; real vegan cheese on the basis of gene-edited yeast; new architectural solutions for physical, biological, economic, and social conditions for successful and productive agricultural solutions in urban environments, for example Plantagon; and new means of transportation which are better for the environment, like Hyperloop. This is where the real challenges lie. Instead of wondering whether we are already living in a computer simulation or arguing about how many angels can dance on the head of a pin, we need to deal with practical, real-life challenges.

Why should we do so? I think the answer has to be a personcentric one. Where they live and their relationship with the environment matters to people. It is not the case that there is a categorical ontological difference between people and the environment, but suffering matters. Using rare earths for digital technologies does not harm the Earth; it has consequences for persons. Forests matter because they are relevant for people, whereby the notion of a person should not be an anthropocentric one. The concept of a person should be a hierarchical one, and should depend on the capacity for suffering of entities, as suffering is morally relevant, whereby we need to develop an empirical means for determining the intensity of suffering of an entity so as to develop a reliable way of determining personhood. This might not be a approach which is satisfactory in all circumstances, but it is an as-good-as-it-gets ethics, and this is all I am trying to present. If the contingent nodal points which we stick to are not plausible anymore, we need to develop new ones.

Political and social institutions are relevant for maintaining a sustainable existence for persons. These are the institutions which are responsible for making policy decisions concerning such critical issues as climate change. Furthermore, personal actions are altered if they move away from a self-understanding that they are the coronation of the Creation, they are the only entities in which God's divine sparks exists. By embracing a humbler self-understanding, on the basis of which we see ourselves as merely gradually different from all other living entities, we can also alter the way we act. There is not just one golden solution by means of which we can deal with tricky global issues such as climate change. As-good-as-it-gets-solutions are what we should aim for. However, a non–dualistic relational understanding of the world definitely supports measures for realizing paradigm-shifts with respect to providing the conditions for a sustainable personal flourishing. We are already on the right track. I can hardly wait for our posthuman future to occur.

Notes

Chapter 1

[1] This chapter is a revised version of the following article of mine (2020b): Transhumanism. In: Thomsen, M.R./Wamberg, J. (eds) *The Bloomsbury Handbook of Posthumanism*. Bloomsbury, London et al, 35–46.

[2] Dante counts as the father of Italian language, who has coined a lot of words. He also wrote the following lines in his text on the paradise:

'Trasumanar significar per verba
non si poria; però l'essemplo basti
a cui esperienza grazia serba.'

(Paradiso, Canto I)

Dante coined the word 'trasumanar', which means to move beyond the limitations of the human. However, this does not turn Dante into a transhumanist. The context in which he used the word make it clear that he did not have a this-worldly evolution in mind.

[3] www.theguardian.com/society/2000/feb/01/futureofthenhs.health (accessed 15 January 2021).

[4] www.aerzteblatt.de/archiv/63471/Rekonstruktion-des-Hymens-Zur-Ethik-eines-tabuisierten-Eingriffs (accessed 29 January 2021).

[5] https://www.forbes.com/sites/robertglatter/2015/11/27/lsd-microdosing-the-new-job-enhancer-in-silicon-valley-and-beyond/?sh=1c561db5188a (accessed 16 June 2021).

[6] https://pubmed.ncbi.nlm.nih.gov/18980888/ (accessed 15 January 2021).

[7] Kevin Warwick realized an initial such implant in 1998 (https://www.wired.com/1998/08/professor-cyborg/ (accessed 16 June 2021)).

[8] https://now.tufts.edu/news-releases/scientists-develop-tiny-tooth-mounted-sensors-can-track-what-you-eat (accessed 16 June 2021).

[9] https://ieet.org/index.php/IEET2/more/hughes20100114/ (accessed 2 September 2020).

[10] https://ourworldindata.org/extreme-poverty (accessed 2 September 2020).

[11] https://ieet.org/index.php/IEET2/cyborgbuddha (accessed 2 September 2020).

[12] www.astrobio.net/news-exclusive/a-cyborg-space-race/ (accessed 2 September 2020).

[13] https://ourworldindata.org/extreme-poverty (accessed 2 September 2020).

[14] The notion of the 'Third Reich' is highly problematic. By using quotation marks, I intend to stress that caution is needed when using the concept.

[15] A different take on transhumanism and utopia has been presented by Bostrom (2008) and Hauskeller (2012).

[16] https://ourworldindata.org/life-expectancy (accessed 2 September 2020).

[17] www.avert.org/professionals/hiv-programming/treatment/overview (accessed 26 August 2020).

[18] https://ourworldindata.org/literacy (accessed 26 August 2020).

[19] https://ourworldindata.org/world-population-growth (accessed 26 August 2020).

20 https://ourworldindata.org/democracy (accessed 26 August 2020).

21 Due to the violence which is connected to any strong concept of truth in correspondence to the world, and the implausibility of such concepts, they ought to be avoided, like speciesism, racism, or sexism, as all of them imply a morally problematic type of discrimination. Alethic nihilism implies the cultural abandonment of alethism, a morally problematic type of discrimination which is related to not affirming the truth in correspondence to the world.

22 On a superficial level, many scholars might agree with the statements concerning alethic and ethical nihilism. However, the consequences of these two attitudes are significant. They have paradigm-shifting implications which become clear if one takes the practical implications into consideration. If you affirm ethical nihilism, the following acts might have to be seen as morally unproblematic: incest among consenting adults; sex with a robot who behaves as if it does not want to be engaged with sexually; selecting a fertilized egg for implantation purposes; selling a kidney (Sorgner 2016e).

Chapter 2

1 This section consists of a revised version of the following article of mine (2020c): Transhumanism without Mind Uploading and Immortality. In: Musiolik, T.H./ Cheok, A.D. (eds) *Analyzing Future Applications of AI, Sensors, and Robotics in Society*. IGI Global, Hershey, PA, 2020, 284–291.

2 www.youtube.com/watch?v=2KK_kzrJPS8 (accessed 1 September 2018).

3 http://news.mit.edu/2017/engineers-3-d-print-living-tattoo-1205 (accessed 1 September 2018).

4 www.independent.co.uk/news/uk/hawking-says-computer-virus-is-form-of-life-susan-watts-on-a-man-made-menace-that-mirrors-its-1374137.html (accessed 1 September 2018).

5 https://chinacopyrightandmedia.wordpress.com/2014/06/14/planning-outline-for-the-construction-of-a-social-credit-system-2014-2020/ (accessed 16 June 2021).

6 Some revised passages from the following article of mine (2019a) have been integrated into some of the following sections: Cyborgs, Freedom, and the Internet Panopticon. In: *Studium Ricerca* 3, 20–44. A detailed treatment of this topic can be found in my monograph *Schöner neuer Mensch* (Sorgner 2018b).

7 www.sciencemag.org/news/2019/01/artificial-intelligence-turns-brain-activity-speech (accessed 9 April 2019).

8 https://www.theguardian.com/technology/2017/apr/19/facebook-mind-reading-technology-f8 (accessed 16 June2021)

9 Some aspects of the following sections contain a further developed and revised version of the following article of mine (2020f): Ein europäisches Sozialkreditsystem als pragmatische Notwendigkeit? In: Bauer, Michael, C./Deinzer, Laura (eds) *Bessere Menschen? Technische und ethische Fragen in der transhumanistischen Zukunft*. Springer, Berlin, Heidelberg, 183–200.

10 https://now.tufts.edu/news-releases/scientists-develop-tiny-tooth-mounted-sensors-can-track-what-you-eat (accessed 9 April 2019).

11 Elsewhere, I have given a detailed analysis of this issue (Sorgner 2018a, 2019c).

12 https://www.southampton.ac.uk/~cpd/history.html (accessed 16 June 2021)

13 https://www.scientificamerican.com/article/fossil-reveals-what-last-common-ancestor-of-humans-and-apes-looked-liked/ (accessed 16 June 2021)

14 www.bbc.co.uk/ethics/animals/using/hybridembryos_1.shtml (accessed 1 January 2021).

15 www.nationalgeographic.com/news/2017/01/human-pig-hybrid-embryo-chimera-organs-health-science/ (accessed 1 January 2021).

16 www.theguardian.com/science/2019/aug/03/first-human-monkey-chimera-raises-concern-among-scientists (accessed 22 February 2020).

17 www.bionews.org.uk/page_135103 (accessed 1 January 2021).

18 These legal arrangements, however, were revoked in 2017: www.newscientist.com/article/2149830-kuwaits-plans-for-mandatory-dna-database-have-been-cancelled/ (accessed 9 April 2019).

19 http://news.mit.edu/2017/engineers-3-d-print-living-tattoo-1205 (accessed 9 April 2019).

20 Some aspects of this issue have been analysed in a different article of mine (Sorgner 2017b). In a recent monograph I have explained how it relates to other aspects of my Nietzschean transhumanism (Sorgner 2019c).

21 https://www.theguardian.com/world/2017/jul/03/colombia-three-men-union-alejandro-rodriguez-manuel-bermudez-victor-hugo-prada (accessed 16 June 2021)

22 https://www.vice.com/en/article/y3madg/people-are-getting-microchipped-in-sweden-and-its-pretty-normal (accessed 16 June 2021).

23 www.cnet.com/news/amazon-go-avoid-discrimination-shopping-commentary/ (accessed 3 January 2021).

24 www.buzzfeednews.com/article/meghara/college-china-social-credit (accessed 22 February 2020).

25 www.youtube.com/watch?v=-Z5WOGSF190 (accessed 11 January 2021). I gave my first TEDx talk at the same event in the Villa Farnesina in Rome in September 2013.

26 www.livescience.com/36284-drug-people-impulsive.html (accessed 3 January 2021).

27 www.healthline.com/health/mental-health/impulse-control#treatments (accessed 3 January 2021).

28 https://papers.ssrn.com/sol3/papers.cfm?abstract_id=2941774 (accessed 23 April 2020).

29 See Willemsen 2016 (my own translations).

30 www.bbc.com/news/blogs-news-from-elsewhere-32050313 (accessed 1 January 2021).

31 Morphological freedom, reproductive freedom, and educational freedom are three central achievements which need to be cherished carefully in a liberal society.

32 Truijens, D./van Exel, J. (2019) Views on Deceased Organ Donation in the Netherlands: A Q-methodology Study. *PLOS One*, 14(5), e0216479. https://doi.org/10.1371/journal.pone.0216479:www.ncbi.nlm.nih.gov/pmc/articles/PMC6534345/#pone.0216479.ref005 (accessed 2 January 2021).

33 https://sparq.stanford.edu/solutions/opt-out-policies-increase-organ-donation (accessed 3 January 2021).

34 This section includes a revised version of reflections from the following article of mine (2020e): Editor's Note. In: *Journal of Posthuman Studies: Philosophy, Media, Technology*, Penn State University Press, 4(2), 113–118.

35 Glocalization is modelled on Japanese 'dochakuka'. See Iwabuchi (2002, 93).

36 The phrase 'was coined in 2006 by Clive Humby, the British mathematician and architect of the Tesco Clubcard, a supermarket reward program'. https://ideas.ted.com/opinion-data-isnt-the-new-oil-its-the-new-nuclear-power/(accessed 27 May 2020).

37 Enago Academy 2018. www.enago.com/academy/china-overtakes-us-with-highest-number-of-scientific-publications/(accessed 19 May 2020).

38 There are voices that claim that the authoritarian use of data may hinder scientific development in the long run (Andrews 2017). However, there are many counterexamples which show that sciences can flourish in authoritarian regimes; for example, Gagarin was the first human being to travel to space. www.history.com/this-day-in-history/first-man-in-space (accessed 22 August 2020).

39 Data analysis is relevant for all aspects of our life world. A list of particularly striking examples can be found at https://processgold.com/blog/top-10-examples-of-successful-data-analysis/ (accessed 18 August 2020). The essence: the more successful the data analysis, the better the quality of life, economic success, and scientific progress.

40 Information about how better data affect the economy can also be found at www.ced.org/blog/entry/big-datas-economic-impact (accessed 18 August 2020).

41 A recent Organization for Economic Co-operation and Development report listed some of the ways that more and better data will affect the economy; see note 35.

42 www.edify.org/from-poverty-to-world-power-koreas-education-story/ (accessed 24 May 2020).

43 www.forbes.com/sites/earlcarr/2020/08/04/reshoring-jobs-to-the-us-versus-made-in-china/#103bcc4c85e1(accessed 23 August 2020).

44 The same applies to Trump's handling of Tik Tok (McCabe 2020). www.faz.net/aktuell/wirtschaft/donald-trump-verlaengert-sanktionen-gegen-huawei-16772240.html (accessed 19 May 2020).

45 www.businessinsider.com/major-us-tech-companies-blocked-from-operating-in-china-2019-5?IR=T (accessed 19 May 2020).

46 www.bbc.com/news/technology-50902496 (accessed 2 January 2021).

47 I explained this in detail in an interview. www.flipsnack.com/BDEAACF6AED/im-apr-may-2020/full-view.html (accessed 24 May 2020).

48 https://news.cgtn.com/news/2020-05-03/German-telecom-carrier-We-need-Huawei-in-5G-network-construction-QcnBZw3ZgA/index.html (accessed 22 May 2020).

Chapter 3

1 This chapter is a revised version of the following article of mine (2016a): From Nietzsche's Overhuman to the Posthuman of Transhumanism: Transcultural Discourses. In: *The Journal of English Language & Literature* 62(2), 163–176.

2 KSA 8, 97: 'Sokrates, um es nur zu bekennen, steht mir so nahe, daß ich fast immer einen Kampf mit ihm kämpfe'.

3 KSA 13, 316f:

> Es gibt keine Übergangsformen. […]
>
> Man behauptet die wachsende Entwicklung der Wesen. Es fehlt jedes Fundament. Jeder Typus hat seine Grenze: über diese hinaus gibt es keine Entwicklung. Bis dahin absolute Regelmäßigkeit. […] Meine Gesamtansicht.– Erster Satz: der Mensch als Gattung ist nicht im Fortschritt. Höhere Typen werden wohl erreicht, aber sie halten sich nicht. Das Niveau der Gattung wird nicht gehoben. Zweiter Satz: der Mensch als Gattung stellt keinen Fortschritt im Vergleich zu irgendeinem andern Tier dar. Die gesamte Tier- und Pflanzenwelt entwickelt sich nicht vom Niederen zum Höheren … Sondern alles zugleich, und übereinander und durcheinander und gegeneinander. Die reichsten und komplexesten Formen – denn mehr besagt das Wort 'höherer Typus' nicht – gehen leichter zugrunde: nur die niedrigsten halten eine scheinbare Unvergänglichkeit fest. Erstere werden selten erreicht und halten sich mit Not oben: letztere haben eine kompromittierende Fruchtbarkeit für sich. – Auch in der Menschheit gehen unter wechselnder Gunst und Ungunst die höheren Typen, die Glücksfälle der Entwicklung, am leichtesten zugrunde. Sie sind jeder Art von décadence ausgesetzt: sie sind extrem, und damit selbst beinahe schon décadents … Die kurze

Dauer der Schönheit, des Genies, des Cäsar ist sui generis: der gleichen vererbt sich nicht. Der Typus vererbt sich; ein Typus ist nichts Extremes, kein 'Glücksfall' … Das liegt an keinem besonderen Verhängnis und 'bösen Willen' der Natur, sondern einfach am Begriff 'höherer Typus': der höhere Typus stellt eine unvergleichlich größere Komplexität – eine größere Summe koordinierter Elemente dar: damit wird auch die Disgregation unvergleichlich wahrscheinlicher. Das 'Genie' ist die sublimste Maschine, die es gibt – folglich die zerbrechlichste. Dritter Satz: die Domestikation (die 'Kultur') des Menschen geht nicht tief … Wo sie tief geht, ist sie sofort die Degenereszenz (Typus: der Christ). Der 'wilde' Mensch (oder, moralisch ausgedrückt: der böse Mensch) ist eine Rückkehr zur Natur – und, in gewissem Sinne, seine Wiederherstellung, seine Heilung von der 'Kultur.'

4 KSA 11, 289: 'Der höchste Mensch würde die größte Vielheit der Triebe haben, und auch in der relativ größten Stärke, die sich noch ertragen läßt. In der Tat: wo die Pflanze Mensch sich stark zeigt, findet man die mächtig gegeneinander treibenden Instinkte (z.B. Shakespeare), aber gebändigt.'

5 KSA 4, 178: 'Menschen, denen es an allem fehlt, außer, daß sie eins zuviel haben – Menschen, welche nichts weiter sind, als ein großes Auge oder ein großes Maul oder ein großer Bauch oder irgend etwas Großes – umgekehrte Krüppel heiße ich solche.'

6 KSA 4, 178: 'Diesem fehlt ein Auge und jenem ein Ohr und einem dritten das Bein, und andre gibt es, die verloren die Zunge oder die Nase oder den Kopf'

7 KSA 4, 19: 'Ich sage euch: man muß noch Chaos in sich haben, um einen tanzenden Stern gebären zu können. Ich sage euch: ihr habt noch Chaos in euch'

8 Nietzsche KSA, GD, 6, 89: '[E]in wohlgeratner Mensch, ein 'Glücklicher', muß gewisse Handlungen tun und scheut sich instinktiv vor andren Handlungen, er trägt die Ordnung, die er physiologisch darstellt, in seine Beziehungen zu Menschen und Dingen hinein.'

9 www.dailymail.co.uk/sciencetech/article-2017818/Embryos-involving-genes-animals-mixed-humans-produced-secretively-past-years.html (accessed 26 April 2020).

10 www.robindelange.com/solar-fish-lustrum-event-leiden-university/ (accessed 26 April 2020).

11 There is plausible scientific evidence that the issue of overpopulation does not have to be a crucial one. I am particularly referring to the work by the economist Max Roser of the University of Oxford and the research published on the following website: https://ourworldindata.org/ (29 January 2021). Some research from the United Nations even suggests that the 12-billionth human being will never be born at all: www.youtube.com/watch?v=QsBT5EQt348&t=3s (accessed 29 January 2021).

12 www.theguardian.com/world/2017/jul/03/colombia-three-men-union-alejandro-rodriguez-manuel-bermudez-victor-hugo-prada (accessed 2 January 2021).

13 This section is a revised version of the following article of mine (2016d): The Stoic Sage 3.0: A Realistic Goal of Moral (Bio)Enhancement Supporters? In: *Journal of Evolution and Technology* 26(1), 83–93.

14 To promote prosocial behaviour by means of oxytocin might also be a promising option, but it faces challenges similar to those faced by the version of moral enhancement I am focusing on at this point. I will return to this. The biotechnological promotion of respect for – and of acts in accordance with –the norms of freedom and equality would be an extremely complex goal. I do not expect it to become a relevant option soon.

15 In order to rid themselves of fear, guilt, and shame, many rapists would be happy to eliminate their sexual and violent compulsions. There are, however, many who would not freely choose this option.

16 I am not implying that any act of religious self-control is an inauthentic crime against our 'true nature'. No such conception of a unified human nature is hidden, or implicit, in my analysis. I merely hold that it seems impossible to make a non-formal statement concerning values that lead to a good life or eudaimonia, because the psychophysiology of human beings differs so radically among the human population that each individual has his or her own idiosyncratic and specific needs, drives, and wishes for living a good life (see Sorgner 2013b). This view is not in conflict with the position that the norms of freedom and equality are legal and moral achievements that human beings have historically fought for. These norms are the result of struggles between various interest groups during and since the Enlightenment. They are not metaphysically valid norms, yet they are wonderful achievements that I wholeheartedly praise and for which I fight, and I am happy that many people agree with me in this respect (see Sorgner 2010a).

17 In fairness to them, their understanding of morality is obviously a wider one than mine. However, it is not relevant in the present context to deal with this issue in any detail.

18 It is interesting that Kant in his anthropology also talked about a historical process, determined by various conflicts, that leads to human moral perfection. His position is founded upon a strange, teleological understanding of historical developments, but his observation that conflict can lead to moral improvement is at least worth noting in this context. See Kant 2006, Part 2.

19 Aristotle, *Pol.* 1 4, 1254a13–17.

20 I stress a correlation between cognitive and moral advances on a social level, since the rejection by Persson and Savulescu of any significant moral progress in our cultural history is concerned with that level. The relationship between the cognitive and moral states of an individual is not a necessary one, that is, it is not impossible that cognitively advanced human beings are morally corrupt. My point relates to a probable social phenomenon: there is a high likelihood that a society with cognitively advanced human beings will also be a society that is morally advanced.

21 An additional remark: in order to critically analyse the question whether there has been 'moral improvement since the time of Confucius, Buddha and Socrates', it might be necessary to consider some further, including more meta-ethical, issues. When discussing morality, Savulescu and Persson refer to a great variety of phenomena, for example non-harming, pro-social behaviour, and altruism. Some of these phenomena suggest that morality implies a certain sort of action, and thus an act-centred understanding of morality (non-harming). At other points, they seem to associate morality with a certain emotional tendency or type of character (altruism), implying an agent-centred or even a virtue-ethical approach. I cannot go into more detail here concerning this highly important issue.

22 This section is a revised version of the following article of mine (2015b): Genetic Enhancement and Metahumanities. The Future of Education. In: *Journal of Evolution and Technology* 25(1), 31–48.

23 By genetic enhancement, in this chapter, I am referring to genetic enhancement by modification, but not to genetic enhancement by selection, for example by selecting fertilized eggs after IVF and PGD.

24 Sections 3.3.1, 3.3.2 and 3.3.7 contain further developed passages from an earlier article of mine (Sorgner 2010b, 4–6).

25 If genetic alterations were irreversible, were made in the interest of children, and were actually in the best interest of the child in most cases, then it could be seen as good that they are irreversible. However, this is not a line of thought that I will consider here.

26 There are some promising examples, for example www.technologyreview.com/2018/11/25/138962/exclusive-chinese-scientists-are-creating-crispr-babies/ (accessed 27 April 2020) I regard the gene scissor CRISPR or gene editing in general is the most important scientific inventions of the beginning of the 21st century.

27 Aristotle, *Nicomachean Ethics*, 1103a.

28 Aristotle, *Categoriae* 8, 8b27–35

29 Aristotle, *Nicomachean Ethics*, III 7, 1114a19–21

30 Helpful complementary arguments concerning this issue can be found in chapter 4 of Blackford's monograph *Humanity Enhanced* (Blackford 2014, ch 4).

31 Maybe, one should rather focus on child abuse, and clarify what counts as abuse, rather than focus on what counts as an all-purpose good. Must education only promote all-purpose goods? Is playing the violin an all-purpose good? If you teach your daughter to play the violin, do you treat her merely as a means/do you merely instrumentalize her?

32 It should be noted that most human beings are interested in an extended health span, but not merely in an extended life span. I have dealt with this issue in more detail in my monograph *Übermensch* (2019c).

33 'Heritable variation – genetic, epigenetic, behavioural, and symbolic – is the consequence both of accidents and of instructive processes during the development' (Japlonka and Lamb 2005, 356). A striking case is that of the evolution of language: 'Dor and Japlonka see the evolution of language as the outcome of the continuous interactions between the cultural and the genetic inheritance system' (Japlonka/Lamb 2005, 307).

34 'Waddington's experiments showed that when variation is revealed by an environmental stress, selection for an induced phenotype leads first to that phenotype being induced more frequently, and then to its production in the absence of the inducing agent' (Japlonka/Lamb 2005, 273).

35 Jonathan M. Levenson and J. David Sweatt show that epigenetic mechanisms probably have an important role in synaptic plasticity and memory formation (2005, 108–118).

36 This section is a revised version of the following article of mine (2014b): Is there a 'Moral Obligation to Create Children with the Best Chance of the Best Life'? In: *Humana Mente: Journal of Philosophical Studies* 26, 199–212.

37 I would not dare to claim to know what a good life is, and I think that there are good reasons for doubting that any non-formal account of goodness is bound to be highly implausible, because any account of the good is closely connected to personal physiopsychological wishes, drives, and desires. This is the reason why I regard the struggles in favour of the norm of negative freedom which have taken place during the Enlightenment as praiseworthy (Sorgner 2010a, 240).

Chapter 4

1 This chapter is a revised version of the following article of mine (2016f): Nietzsche's Virtue Ethics and Sandels' Rejection of Enhancement Technologies: Truthful, Virtuous Parents May Enhance their Children Genetically. In: Deretić, Irina/Sorgner, Stefan Lorenz (eds) *From Humanism to Meta- Post and Transhumanism?* Peter Lang, New York et al, 359–378.

2 It has to be noted that it is anachronistic to refer to Nietzsche as a communitarian thinker. Historically, philosophical communitarianism was developed as a reaction to the liberal political philosophies which were put forward during the 1970s Harvard University by Rawls and Nozick. However, in this section I will use the concept 'communitarianism' in an ahistorical sense, whereby I focus on some central characteristics which are associated with the concept. Communitarianism regards the good as prior to the right. Liberalism sees

the right as prior to the good. The question of the good seeks universally valid conditions for all human beings to live a good life. The question of the right, on the other hand, searches for moral laws which are universally valid, but without their being necessarily connected to the good. Liberals tend to doubt the possibility that the content of the concept of the good can be determined, because they uphold that human beings differ so greatly from one another that a plurality of goods needs to be enabled within a social organization. Communitarianists, on the other hand, uphold that meaningful judgments concerning the concept of the good can be made and that these ought to provide the value basis of a social structure because they enable all the citizens of the respective community to lead good and flourishing lives. Hence, liberals and communitarians differ concerning the descriptive and normative content of their anthropologies

3   Eugenics refers to any type of technological method to alter the genetic make-up of human beings (Sorgner 2006, 201–209). Liberal eugenics, in contrast to state-imposed methods of eugenics, stresses that the decision to implement an alteration may be made either by oneself or by the parents for their children. Once a different institution like the state has the right to make such a decision, then a state-regulated type of eugenics is in force, which was present in many countries about a century ago, and a particular, cruel and organized version of this type of eugenics was carried through during the 'Third Reich'. Most Western people agree that the state-governed version of eugenics is morally condemnable, and I agree. At the end of twentieth century, academic debates arose concerning the legitimacy of liberal eugenics, due to rapid developments in the field of human biotechnologies. Later, many bioethicists decided to talk about genetic enhancement rather than liberal eugenics in order to avoid the morally charged word 'eugenics'. Nowadays, there is a tendency for bioconservatives employ the term 'liberal eugenics', but bioliberals tend to use the concept 'genetic enhancement'. In both cases this is presumably due to the emotional content and the connotations which are associated with the concepts in question.

4   Leading transhumanists, bioethicists, and Nietzsche scholars participated in the debate (Sorgner 2009, 29–42; Hauskeller 2010, 5–8; Hibbard 2010, 9–12; More 2010, 1–4; Sorgner 2010a, 1–19; Sorgner 2011b, 1–46; Ansell-Pearson 2011, 1–16; Babich 2011, 1–39; Loeb 2011, 1–29. In the meantime, the debate has already been received in the Nietzsche secondary literature, for example Woodward 2011. Concerning a related topic of the debate, see also Sorgner 2013a. Central articles of the Nietzsche and transhumanism debate have been published in the essay collection 'Nietzsche and Transhumanism', which was edited by Yunus Tuncel (2017).

5   In an article from 2014, Habermas relates transhumanism to the way of thinking of a sect (Habermas 2014, 36). It was his rhetorical attempt to exclude transhumanism from morally acceptable public and academic discourses. His former understanding that transhumanism is shared only among a small group of 'freaked out individuals' was no longer valid in 2014, as it had already entered various cultural spheres at that time (in particular via movies, TV series, novels). I will deal with the great variety of cultural aspects of transhumanism in a separate monograph.

6   I have already mentioned one of his arguments against autonomous genetic enhancement earlier in this section. Sandel's point of view implies that currently dominant moral principles like autonomy or equality do not imply that genetic enhancement has to be morally bad. However, on the basis of Sandel's virtue-ethical approach, anyone who uses genetic enhancement is bound to have a vicious character. To analyse the plausibility of this general rejection further, it is relevant to consider in detail which specific consequences are implicit in the aforementioned line of argument.

7   For a more detailed argument in favour of this structural analogy, see Sorgner 2010a, section 1.

8   See section 8 of the following article of mine: Sorgner (2011b): Zarathustra 2.0 and Beyond: Further Remarks on the Complex Relationship between Nietzsche and Transhumanism. In: *The Agonist. A Nietzsche Circle Journal* 4(2), 1–46.

9   This section is a revised version of the following article of mine (2016b): Three Transhumanist Types of (Post)Human Perfection. In: Hurlbut, J.B./Tirosh-Samuelson, H. (eds) *Perfecting Human Futures: Transhuman Visions and Technological Imaginations*. Springer, Wiesbaden, 141–157.

10 I regularly need to stress that freedom is an amazing achievement. Freedom means negative freedom, the absence of constraint, and includes morphological freedom, educational freedom, and reproductive freedom.

11 This section is a revised version of the following article of mine (2020a): What Does It Mean to Harm a Person? In: *Humana Mente. Journal of Philosophical Studies* 13(37), 207–232.

12 This section includes revised sections of the following article of mine (2013d): Kant, Nietzsche and the Moral Prohibition of Treating a Person Solely as a Means. In: *The Agonist. A Nietzsche Circle Journal* 4(1–2), 1–6.

13 www.bpb.de/nachschlagen/zahlen-und-fakten/soziale-situation-in-deutschland/145148/religion (accessed 29 January 2021).

14 https://www.scientificamerican.com/article/do-video-games-inspire-violent-behavior/ (accessed 21 June 2021).

15 www.scientificamerican.com/article/argentina-grants-an-orangutan-human-like-rights/ (accessed 15 March 2019).

16 In the following passages, I criticize that self-consciousness ought to be seen as necessary for personhood. In my book *Schöne neue Welt* (2018b, 13), I explain why sentience is not necessary for personhood either.

17 www.animalcognition.org/2015/04/15/list-of-animals-that-have-passed-the-mirror-test/ (accessed 23 April 2020).

18 www.buzzfeednews.com/article/rosalindadams/aibo-robot-dogs-japan (accessed 23 April 2020).

19 www.theinvisiblegorilla.com/gorilla_experiment.html (accessed 23 April 2020).

20 https://ghr.nlm.nih.gov/condition/congenital-insensitivity-to-pain (accessed 23 April 2020).

21 www.ncbi.nlm.nih.gov/pubmed/23374102 (accessed 23 April 2020).

22 www.bbc.com/news/world-europe-43839655 (accessed 23 April 2020).

23 www.gesetze-im-internet.de/stgb/__218.html (accessed 23 April 2020). Only if a woman became pregnant due to a rape, or if the life of a pregnant woman is threatened, an abortion is not illegal in Germany.

24 www.lifetimedaily.com/want-live-forever/ (accessed 23 April 2020).

25 www.nytimes.com/2020/03/01/business/china-coronavirus-surveillance.html (accessed 2 January 2021).

26 This section is a revised version of the following article of mine (2018c): Transhumanism and the Land of Cockaygne. In: *Trans-Humanities* 11(1), 165–188.

27 www.vice.com/en_us/article/8q854v/elon-musk-simulated-universe-hypothesis (accessed 1 September 2020).

28 https://interestingengineering.com/science-or-scam-9-things-you-should-know-before-you-sign-up-for-cryonic-suspension (accessed 10 August 2020).

29   www.independent.co.uk/news/science/scientists-have-successfully-revived-an-animal-frozen-30-years-ago-a6816656.html (accessed 10 August 2020), www.natureworldnews.com/articles/19877/20160211/cryogenics-entire-rabbit-brain-successfully-frozen-revived-first-time.htm (accessed 10 August 2020).

30   https://www.ncbi.nlm.nih.gov/books/NBK545144/ (accessed 19 June 2021).

31   www.independent.co.uk/news/uk/hawking-says-computer-virus-is-form-of-life-susan-watts-on-a-man-made-menace-that-mirrors-its-1374137.html (accessed 1 September 2018).

32   https://de.wikisource.org/wiki/Faust_-_Der_Trag%C3%B6die_zweiter_Teil 1 September 2020, lines 11580-11585 (accessed 3 January 2021).

References

Agar, N. (2004) *Liberal Eugenics: In Defence of Human Enhancement.* Blackwell, Oxford.

Agganyani, V. (2013) Dukkha: Suffering. In: Runehov, A.L.C./Oviedo, L. (eds) *Encyclopedia of Sciences and Religions.* Springer, Dordrecht. https://doi.org/10.1007/978-1-4020-8265-8_201206

Aknin, L.B./Broesch, T./Hamlin, J.K./Van de Vondervoort, J.W. (2015) Prosocial Behavior Leads to Happiness in a Small-scale Rural Society. In: *Journal of Experimental Psychology* 144, 788–795.

Amsterdam, B. (1972) Mirror Self-image Reactions before Age Two. In: *Developmental Psychobiology* 5(4), 297–305.

Ansell-Pearson, K. (2011) The Future is Superhuman. Nietzsche's Gift. In: *The Agonist. A Nietzsche Circle Journal* 4(2), 1–16.

Auweele, Dennis Vanden (2017) *The Kantian Foundation of Schopenhauer's Pessimism.* Routledge, London.

Babich, B. (2011) On the 'All-too-Human' Dream of Transhumanism. In: *The Agonist. A Nietzsche Circle Journal* 4(2), 1–39.

Bacon, F. (1859) Meditationes sacrae. In: Spedding, J./Leslie Ellis, R./Denon Heath, D. (eds) *Francis Bacon (1857–1874) The Works of Francis Bacon,* vol VII. Longman, London, 227–254.

Badmington, N. (2000) Approaching Posthumanism. In: Badmington, N. (ed) *Posthumanism.* Palgrave Macmillan, New York, 1–10.

Beese, F. (2004) *Was ist Psychotherapie? Ein Leitfaden für Laien zur Information über ambulante und stationäre Psychotherapie.* V&R, Goettingen.

Birch, K. (2005) Beneficence, Determinism and Justice. An Engagement with the Argument for the Genetic Selection of Intelligence. In: *Bioethics* 19(1), 12–28.

Blackford, R. (2014) *Humanity Enhanced. Genetic Choice and the Challenge for Liberal Democracies.* MIT Press, Cambridge, MA.

Blackford, R. (2017) Nietzsche, the Übermensch, and Transhumanism. Philosophical Reflections. In: Tuncel, Y. (ed) *Nietzsche and Transhumanism. Precursor or Enemy.* Cambridge Scholars Press, Newcastle upon Tyne, 191–204.

Bostrom, N. (2001) Transhumanist Values. Version of 18 April 2001 (accessed 14 December 2018) www.nickbostrom.com/tra/values.html

Bostrom, N. (2003) Are You Living in a Computer Simulation? In: *Philosophical Quarterly* 53(211), 243–255.

Bostrom, N. (2005) A History of Transhumanist Thought. In: *Journal of Evolution and Technology* 14(1), 1–25.

Bostrom, N. (2008) Letter from Utopia. In: *Studies in Ethics, Law, and Technology*, 2(1), 1–7.

Bostrom, N. (2009) Why I Want to Be a Posthuman when I Grow Up. In: Gordijn, B./Chadwick, R. (eds) *Medical Enhancement and Posthumanity*. Springer, New York, 107–136.

Buddensiek, F. (2002) Hexis. In: Horn, C./Rapp, C. (eds) *Wörterbuch der antiken Philosophie*. Beck, Munich.

Cabat-Zinn, J. (2017) Foreword. In: Varela, F.J./Thompson, E./Rosch, E. (eds) *The Embodied Mind*. MIT Press, London et al.

Capek, M. (1991) *The New Aspects of Time. Its Continuity and Novelties.* Kluwer, Dordrecht et al.

Caplan, R. (2012) *Selfish Reasons to Have More Kids. Why Being a Great Parent is Less Work and More Fun Than You Think.* Basic Books, New York.

Cazzola Gatti, R. (2015) Self-consciousness. Beyond the Looking-glass and What Dogs Found There. In: *Ethology Ecology & Evolution* 28(2), 232–240.

Crockett, M.J./Clark, L./Hauser, M.D./Robbins, T.W. (2010a) Serotonin Selectively Influences Moral Judgment and Behavior through Effects on Harm Aversion. In: *Proceedings of the National Academy of Sciences U.S.A.* 107(40), 17433–17438.

Crockett, M.J./Clark, L./Hauser, M.D./Robbins, T.W. (2010b) Reply to Harris and Chan. Moral Judgment is more than Rational Deliberation. In: *Proceedings of the National Academy of Sciences U.S.A.* 107(50), E184.

De Chardin, Teilhard ([1959] 2004) *The Future of Man*. Doubleday, New York.

De Grey, A. (2007) Ending Aging. *The Rejuvenation Breakthroughs that Could Reverse Aging in Our Lifetime.* St. Martin's Griffin, New York.

De Grey, A./Rae, M. (2010) Never Old! This Is How Aging Can Be Reversed. Progress in Rejuvenation Research. Transcript, Bielefeld.

De Melo-Martin, I. (2004) On Our Obligation to Select the Best Children. A Reply to Savulescu. In: *Bioethics* 18(1), 72–83.

Del Val, J./Sorgner, S.L. (2011) A Metahumanist Manifesto. In: *The Agonist. A Nietzsche Circle Journal* 4(2), 1–4.

Douglas, T. (2011) Moral Enhancement. In: Savulescu, J./ter Meulen, J./Kahane, G. (eds) *Enhancing Human Capacities*. Wiley-Blackwell, Malden, MA, 467–485.

Doyle, D.J. (2018) *What Does it Mean to be Human? Life, Death, Personhood and the Transhumanist Movement.* Springer, Cham.

Eames, S.M. (1977) *Pragmatic Naturalism. An Introduction.* Southern Illinois University Press, Carbondale, IL.

Ehni, Hans-Jörg (ed) (2018) *Altersutopien.* Campus, Frankfurt a.M./ New York.

Eissa, T.L./Sorgner, S.L. (eds) (2011) *Geschichte der Bioethik.* Mentis, Paderborn.

Esfandiary, F.M. (1973) *Up-Wingers. A Futurist Manifesto.* John Day Co., New York.

Esfandiary, F.M. (1974) Transhumans-2000. In: Tripp, M. (ed) *Woman in the Year 2000.* Arbor House, New York, 291–298.

Ettinger, R.C.W. ([1972] 2005) *Man into Superman.* Ria University Press, Palo Alto.

Floridi, L. (2014) *The 4th Revolution. How the Infosphere Is Reshaping Human Reality.* Oxford University Press, Oxford.

Flynn, J.R. (2012) *Are We Getting Smarter? Rising IQ in the Twenty-first Century.* Cambridge University Press, Cambridge.

FM-2030 (1989) *Are You a Transhuman?* Warner Books, New York.

Foucault, M. (1977) *Discipline and Punish. The Birth of the Prison.* Translated by A. Sheridan. Pantheon, New York.

Fukuyama, F. (2004) The World's Most Dangerous Ideas: Transhumanism. In: *Foreign Policy* 144(Sept/Oct), 42–43.

Goethe, J.W. (2008) *Faust, Part One.* Translated by D. Luke. Oxford University Press, Oxford.

Grof, S. (1975) *Realism of the Human Unconscious. Observations from LSD Research.* Souvenir Press, London

Güth, W./Schmittberger, R./Schwarze, B. (1982) An Experimental Analysis of Ultimatum Bargaining. In: *Journal of Economic Behavior and Organization* 3, 367–388.

Habermas, J. (2001) *Die Zukunft der menschlichen Natur. Auf dem Weg zu einer liberalen Eugenik?* Suhrkamp, Frankfurt am Main.

Habermas, J. (2003) *The Future of Human Nature.* Translated by H. Beister, M. Pensky, W. Rehg. Cambridge, Polity Press.

Habermas, J. (2004) Freiheit und Determinismus. In: *Deutsche Zeitschrift für Philosophie* 52(6), 871–890.

Habermas, J. (2014) Autopoietische Selbsttransformationen der Menschengattung. In: Habermas, J./Ehalt, H.C./Körtner, U.H.J./Kampits, P., *Biologie und Biotechnologie–Diskurse über eine Optimierung des Menschen.* Picus Verlag, Wien, 27–37.

Harari, Y.N. (2020) The World after Coronavirus. In: *Financial Times*, 20 March (accessed 5 April 2020) www.ft.com/content/ 19d90308-6858-11ea-a3c9-1fe6fedcca75

Haraway, D. (1991) A Cyborg Manifesto: Science, Technology, and Socialist-Feminism in the Late Twentieth Century. In: Haraway, D., *Simians, Cyborgs and Women: The Reinvention of Nature*. Routledge, New York, 149–181.

Harris, J. (2011) Moral Enhancement and Freedom. In: *Bioethics* 25(2), 102–111.

Harris, J./Chan, S. (2010) Moral Behavior Is not What It Seems. In: *Proceedings of the National Academy of Sciences U.S.A.* 107(50), E183.

Hauskeller, M. (2010) Nietzsche, the Overhuman, and the Posthuman. A Reply to Stefan Sorgner. In: *Journal of Evolution and Technology* 21(1), 5–8.

Hauskeller, M. (2012) Reinventing Cockayne. Utopian Themes in Transhumanist Thought. In: *Hastings Center Report* 42(2), 39–47.

Hayles, N. Katherine (1999) *How We Became Posthuman. Virtual Bodies in Cybernetics, Literature, and Informatics*. University of Chicago Press, Chicago.

Hayles, N. Katherine (2017) *Unthought. The Power of the Cognitive Nonconscious*. University of Chicago Press, Chicago.

Herissone-Kelly, P. (2006) Procreative Beneficence and the Prospective Parent. In: *Journal of Medical Ethics* 32, 166–169.

Hibbard, B. (2010) Nietzsche's Overhuman Is an Ideal Whereas Posthumans Will Be Real. In: *Journal of Evolution and Technology* 21(1), 9–12.

Hoerster, N. (2002) *Ethik des Embryonenschutzes. Ein rechtsphilosophischer Essay*. Reclam, Stuttgart.

Hoerster, N. (2013) *Wie schutzwürdig ist der Embryo? Zur Abtreibung, PID und Embryonenforschung*. Velbrück Wissenschaft, Weilerswist.

Hughes, J. (2004) *Citizen Cyborg. Why Democratic Societies Must Respond to the Redesigned Human of the Future*. Westview Press, Boulder.

Hughes, J. (2014) Politics. In: Ranisch, R./Sorgner, S.L. (eds) *Post- and Transhumanism: An Introduction*. Peter Lang, Frankfurt am Main, 133–148.

Huxley, J. (1951) Knowledge, Morality, and Destiny. The William Alanson White Memorial Lectures, 3rd series. *Psychiatry* 14(2), 127–151.

Huxley, J. (1957) *New Bottles for New Wine*. Chatto & Windus, London.

Istvan, Z. (2013) *The Transhumanist Wager*. Futurity Imagine Media LLC, Reno.

Iwabuchi, K. (2002) *Recentering Globalization. Popular Culture and Japanese Transnationalism*. Duke University Press, Durham, NC/London.

Japlonka, E./Lamb, M.J. (2005) *Evolution in Four Dimensions. Genetic, Epigenetic, Behavioral, and* symbolic variation *in the History of Life*. MIT Press, Cambridge, MA.

Kahn, C.H. (2008) *The Art and Thought of Heraclitus: An Edition of the Fragments with Translation and Commentary*. Cambridge University Press, Cambridge et al.

Kant, I. (1902–) *Gesammelte Schriften* in 29 vols. Akademieausgabe, Berlin et al.

Kant, I. (2006) *Anthropology from a Pragmatic Point of View*. Edited by R.B. Louden and M. Kuehn. Cambridge University Press, Cambridge.

Kind, A. (2020) *Philosophy of Mind: The Basics*. Routledge, London.

Knoepffler, N./Schipanski, D./Sorgner, S.L. (eds) (2007) *Human Biotechnology as Social Challenge*. Ashgate, Aldershot.

Koechy, K. (2006) Gentechnische Manipulation und die Naturwuechsichkeit des Menschen. Bemerkungen zu Habermas. In: Sorgner, S.L./Birx, H. James/Knoepffler, Nikolaus (eds) *Eugenik und die Zukunft*. Alber Verlag, Freiburg i. B., 71–84.

Kurzweil, R. (2006) *The Singularity Is Near. When Humans Transcend Biology*. Penguin, London.

Kurzweil, R./Grossman, T. (2011) *Transcend. Nine Steps to Living Well Forever*. Rodale Books, New York.

Levenson, J.M./Sweat, J.D. (2005) Epigenetic Mechanisms in Memory Formation. In: *Nature Reviews, Neuroscience* 6(2), 108–118.

Liu, N.Y. (2012) *Bio-privacy. Privacy Regulations and the Challenge of Biometrics*. Routledge, London.

Loeb, P.S. (2011) Nietzsche's Transhumanism. In: *The Agonist. A Nietzsche Circle Journal* 4(2), 1–29.

Maguire, A.M./Simonelli, F./Pierce, E.A./Pugh Jr, E.N./Mingozzi, F./ Bennicelli, J./Banfi, S./Marshall, K.A./Testa, F./Surace, E.M./Rossi, S./ Lyubarsky, A./Arruda, V.R./Konkle, B./Stone, E./Sun, J./Jacobs, J./ Dell'Osso, L./Hertle, R./Ma, J./Redmond, T.M./Zhu, X./Hauck, B./ Zelenaia, O./Shindler, K.S./Maguire, M.G./Wright, J.F./Volpe, N.J./ Wellman McDonnell, J./Auricchio, A./High, K.A./Bennett, J. (2008) Safety and Efficacy of Gene Transfer for Leber's Congenital Amaurosis. In: *New England Journal of Medicine* 358, 2240–2248.

Malcolm, J. (1984) *In the Freud Archives*. Jonathan Cape, London.

Margalit, A. (2012) *Politik der Würde*. Suhrkamp, Berlin.

McCaffrey, A.P./Meuse, L./Pham, T.T./Conklin, D.S./Hannon, G.J./Kay, M.A. (2002) RNA Interference in Adult Mice. In: *Nature* 418(6893), 38–39.

Millon, T. (2011) *Disorders of Personality. Introducing a DSM/ICD Spectrum from Normal to Abnormal*. John Wiley & Sons, Hoboken, NJ.

More, M. (1990) Transhumanism. Toward a Futurist Philosophy. In: *Extropy* (6), 6–12.

More, M. (1993) Technological Self-Transformation. Expanding Personal Extropy. In: *Extropy* (10), 15–24.

More, M. (2010) The Overhuman in the Transhuman. In: *Journal of Evolution and Technology* 21(1), 1–4.

More, M. (2013) The Philosophy of Transhumanism. In: More, M./ Vita-More, N. (eds) *The Transhumanist Reader: Classical and Contemporary Essays on the Science, Technology, and Philosophy of the Human Future.* Wiley-Blackwell, Chichester, 3–17.

More, M./ Vita-More, N. (2013) (eds) *The Transhumanist Reader. Classical and Contemporary Essays on the Science, Technology, and Philosophy of the Human Future.* Wiley-Blackwell, Chichester.

Morris, K.V./Chan, S.W.-L./Jacobsen, S.E./Looney, D.J. (2004) Small Interfering RNA-induced Transcriptional Gene Silencing in Human Cells. In: *Science* 305(5688), 1289–1292.

Niemeyer, C. (2016) *Nietzsche als Erzieher. Pädagogische Lektüren und Relektüren.* Juventa, Weinheim & Basel.

Nietzsche, F. (1967ff) *Sämtliche Werke. Kritische Studienausgabe in 15 Bänden.* Edited by G. Colli and M. Montinari. München/New York: Deutscher Taschenbuch Verlag.

Nietzsche, F. (1968) *The Will to Power.* Translated by W. Kaufmann and R.J. Hollingdale. Weidenfeld and Nicolson, London.

Olson, D.R. (2003) *Psychological Theory and Educational Reform. How School Remakes Mind and Society.* Cambridge University Press, Cambridge.

Ottaway, A.K.C. (1999) *Education and Society. An Introduction to the Sociology of Education.* Routledge, London.

Parker, M. (2007) The Best Possible Child. In: *Journal of Medical Ethics* 33(5), 279–283.

Parkes, G. (ed) (1996) *Nietzsche and Asian Thought.* Chicago University Press, Chicago.

Pearce, D. (1995) The Hedonist Imperative. HedWeb, 1995 (accessed 10 August 2021) https://www.hedweb.com/confile.htm

Persson, I./Savulescu, J. (2008) The Perils of Cognitive Enhancement and the Urgent Imperative to Enhance the Moral Character of Humanity. In: *Journal of Applied Philosophy* 25, 162–177.

Persson, I./Savulescu, J. (2012) *Unfit for the Future. The Need for Moral Enhancement.* Oxford University Press, Oxford.

Persson, I./Savulescu, J. (2013) Getting Moral Enhancement Right. The Desirability of Moral Bioenhancement. *Bioethics* 27(3), 124–131.

Pinker, S. (2011) *The Better Angels of Our Nature. Why Violence Has Declined.* Viking Books, New York.

Rabelhofer, B. (2006) *Symptom, Sexualitaet, Trauma.* K&N, Wuerzburg.

Raju, C.K. (2003) *The Eleven Pictures of Time: The Physics, Philosophy, and Politics of Time Beliefs.* Sage, Thousand Oaks, CA.

Rentrop, M./Müller, R./Baeuml, J. (2009) *Klinikleitfaden Psychiatrie und Psychotherapie.* Elsevier, Munich.

Rollins, M.D./Rosen, M.A. (2012) Gregory's Pediatric Anesthesia. In: Gregory, G.A./Andopoulos, D.B. (eds) *Anesthesia for Fetal Intervention and Surgery*. Blackwell, Oxford, 444–474

Rosa, H. (2014) *Entschleunigung. Die Veränderung der Zeitstrukturen in der Moderne*. Suhrkamp, Frankfurt a. M.

Rothblatt, M. (2015) *Virtually Human. The Promise and the Peril of Digital Immortality*. Picador, London et al.

Ryan, C. (2017) Schopenhauer and Gotama on Life's Suffering. In: Shapshay, S. (ed) *The Palgrave Schopenhauer Handbook. Palgrave Handbooks in German Idealism*. Palgrave Macmillan, Cham.

Sambursky, S. (1976) The Stoic Doctrine of Eternal Recurrence. In: Capek, M. (ed) *The Concepts of Space and Time: Their Structure and Their Development*. Reidel Publishing, Dordrecht/Boston, 167–173.

Sampanikou, E. D./Stasienko, J. (eds.) (2021) *Posthuman Studies Reader. Core readings on Transhumanism, Posthumanism and Metahumanism*. Schwabe, Basel.

Sandberg, A. (2013) Morphological Freedom. Why We Not Just Want It, but Need It. In: More, M./Vita-More, N. (2013) (eds) *The Transhumanist Reader. Classical and Contemporary Essays on the Science, Technology, and Philosophy of the Human Future*. Wiley-Blackwell, Chichester, 56–64.

Sandel, M.J. (2007) *The Case against Perfection. Ethics in the Age of Genetic Engineering*. Harvard University Press, Cambridge, MA.

Savulescu, J. (2001) Procreative Beneficence. Why We Should Select the Best Children. In: *Bioethics* 15(5–6), 413–426.

Savulescu, J./Kahane, G. (2009) The Moral Obligation to Create Children with the Best Chance of the Best Life. In: *Bioethics*, 23(5), 274–290.

Schrauwers, A./Poolmann, B. (2013) *Synthetische Biologie. Der Mensch als Schöpfer*. Springer, Berlin/Heidelberg.

Sellars, R.W. (1922) Evolutionary Naturalism. In: *Nature* 110, 631.

Sender, R./ Fuchs, S./Milo, R. (2016) Revised Estimates for the Number of Human and Bacteria Cells in the Body. In: *PLoS Biology* 14(8), e1002533.

Singer, P. (2002) *Animal Liberation*. Harper, New York.

Singer, P. (2011a) *Practical Ethics*. 3rd edn. Cambridge University Press, Cambridge.

Singer, P. (2011b) Vorwort. In: Eissa, T.-L./Sorgner, S.L. (eds) *Geschichte der Bioethik: Eine Einführung*. Mentis, Paderborn, 13–15.

Sloterdijk, P. (1999) *Regeln für den Menschenpark. Ein Antwortschreiben zu Heideggers Brief über den Humanismus*. Suhrkamp, Frankfurt am Main.

Sloterdijk, P. (2009) Rules for the Human Zoo. A Response to the 'Letter on Humanism'. In: *Environment and Planning D: Society and Space* 27(1), 12–28.

Song, E./Lee, S.-K./Wang, J./Ince, N./Ouyang, N./Min, J./Chen, J./ Shankar, P./Lieberman, J. (2003) RNA Interference Targeting Fas Protects Mice from Fulminant Hepatitis. In: *Nature Medicine* 9, 347–351.

Sorgner, S.L. (2004) Two Concepts of 'Liberal Education'. In: *ethic@* 3(2), 107–119.

Sorgner, S.L. (2006) Facetten der Eugenik. In: Sorgner, S.L./Birx, H. James/ Knoepffler, N. (eds) *Eugenik und die Zukunft*. Alber Verlag, Feiburg i. B., 201–209.

Sorgner, S.L. (2007) *Metaphysics without Truth. On the Importance of Consistency within Nietzsche's Philosophy*. 2nd rev edn. University of Marquette Press, Milwaukee WI.

Sorgner, S.L. (2008) Ethik. In: Sorgner, S.L./Birx, H. James/Knoepffler, N. (eds) *Wagner und Nietzsche: Kultur-Werk-Wirkung: Ein Handbuch*. Rowohlt, Reinbek b. Hamburg, 194–214.

Sorgner, S.L. (2009) Nietzsche, the Overhuman, and Transhumanism. In: *Journal of Evolution and Technology* 21(1), 29–42.

Sorgner, S.L. (2010a) *Menschenwürde nach Nietzsche. Die Geschichte eines Begriffs*. WBG, Darmstadt.

Sorgner, S.L. (2010b) Beyond Humanism. In: *Journal of Evolution and Technology* 21(2), 1–19.

Sorgner, S.L. (2011a) Reflexionen zum Musikdrama. Richard Wagner, Thomas Mann und der Posthumanismus. In: Pils, H./Ulrich, C. (eds) *Liebe ohne Glauben. Thomas Mann und Richard Wagner*. Wallstein Verlag, Göttingen, 152–172.

Sorgner, S.L. (2011b) Zarathustra 2.0 and Beyond: Further Remarks on the Complex Relationship between Nietzsche and Transhumanism. In: *The Agonist. A Nietzsche Circle Journal* 4(2), 1–46.

Sorgner, S.L. (2013) Evolution, Education, and Genetic Enhancement. In: Sorgner, S.L./Jovanovic, B.-R. (eds) *Evolution and the Future. Anthropology, Ethics, Religion*. Peter Lang, Frankfurt a. M. et al, 85–100.

Sorgner, S.L. (2013a) Human Dignity 2.0. Beyond a Rigid Version of Anthropocentrism. In: *Trans-Humanities* 6(1), 135–159.

Sorgner, S.L. (2013b) Paternalistic Cultures versus Nihilistic Cultures. In: *European Journal of Science and Theology* 9(1), 55–60.

Sorgner, S.L. (2013c) Wagners (un)zeitgemäße Betrachtungen. Reaktionäre oder progressive Überlegungen zum Musikdrama? In: Loos, H. (ed) *Richard Wagner. Persönlichkeit, Werk und Wirkung*. Sax Verlag, Markkleeberg, 193–200.

Sorgner, S.L. (2013d) Kant, Nietzsche and the Moral Prohibition of Treating a Person Solely as a Means. In: *The Agonist. A Nietzsche Circle Journal* 4(1–2), 1–6.

Sorgner, S.L. (2014a) Wagners Ideal des Guten als Inspiration für das 21. Jahrhundert. Überlegungen zur Dialektik von Liebe und Macht im Rheingold und im Posthumanismus. In: Friedrich, S. (ed) *Das Kunstwerk der Zukunft*. K&N, Würzburg, 123–137.

Sorgner, S.L. (2014b) Is there a Moral Obligation to Create Children with the Best Chance of the Best Life? In: *Humana Mente: Journal of Philosophical Studies* 26, 199–212.

Sorgner, S.L. (2015a) Metahumanist Politics and Three Types of Freedom. In: *European Journal of Science and Theology* 11(1), 225–236.

Sorgner, S.L. (2015b) Genetic Enhancement and Metahumanities. The Future of Education. In: *Journal of Evolution and Technology* 25(1), 31–48.

Sorgner, S.L. (2016a) From Nietzsche's Overhuman to the Posthuman of Transhumanism. Transcultural Discourses. In: *The Journal of English Language & Literature* 62(2), 163–176.

Sorgner, S.L. (2016b) Three Transhumanist Concepts of Human Perfection: In: Hurlbut, B./Tirosh-Samuelson, H. (eds) *Perfecting Human Futures: Technology, Secularization and Eschatology*. Springer, Wiesbaden, 141–157.

Sorgner, S.L. (2016d) The Stoic Sage 3.0. A Realistic Goal of Moral (Bio) Enhancement Supporters? In: *Journal of Evolution and Technology* 26(1), 83–93.

Sorgner, S.L. (2016e) *Transhumanismus: 'Die gefährlichste Idee der Welt!?'* Herder, Freiburg i. Br.

Sorgner, S.L. (2016f) Nietzsche's Virtue Ethics and Sandels' Rejection of Enhancement Technologies: Truthful, Virtuous Parents may enhance their Children Genetically. In: Deretić, I./Sorgner, S.L. (eds) *From Humanism to Meta- Post and Transhumanism?* Peter Lang, New York et al, 359–378.

Sorgner, S.L. (2017a) Philosophy as 'Intellectual War of Values'. In: Blackford, R./Broderick, D. (eds) *Philosophy's Future. The Problem of Philosophical Progress*. Wiley/Blackwell, Hoboken/NY, 193–200.

Sorgner, S.L. (2017b) Genetic Privacy, Big Gene Data, and the Internet Panopticon. In: *Journal of Posthuman Studies: Philosophy, Media, Technology* 1(1), 87–103.

Sorgner, S.L. (2018a) Altern als Krankheit. Ein Plädoyer für unkonventionelles Denken: In: Ehni, H.-J. (ed) *Altersutopien: Medizinische und gesellschaftliche Zukunftshoffnungen der Lebensphase Alter*. Campus, Frankfurt a. M., 56–74.

Sorgner, S.L. (2018b) *Schöner neuer Mensch*. Nicolai Verlag, Berlin.

Sorgner, S.L. (2018c) Transhumanism and the Land of Cockaygne. In: *Trans-Humanities* 11(1), 165–188.

Sorgner, S.L. (2019a) Cyborgs, Freedom, and the Internet Panopticon. In: *Studium Ricerca* (3), 20–44.

Sorgner, S.L. (2019b) Editor's Note. In: *Journal of Posthuman Studies. Philosophy, Media, Technology*, 3(1), 1–4.

Sorgner, S.L. (2019c) *Übermensch. Plädoyer für einen Nietzscheanischen Transhumanismus*. Schwabe, Basel.

Sorgner, S.L. (2020a) What Does It Mean to Harm a Person? In: *Humana Mente. Journal of Philosophical Studies* 13(37), 207–232.

Sorgner, S.L. (2020b) Transhumanism. In: Thomsen, M.R./Wamberg, J. (eds) *The Bloomsbury Handbook of Posthumanism*. Bloomsbury, London, 35–46.

Sorgner, S.L. (2020c) Transhumanism without Mind Uploading and Immortality. In: Musiolik, T.H./Cheok, A.D. (eds) *Analyzing Future Applications of AI, Sensors, and Robotics in Society*. IGI Global, Hershey, PA, 284–291.

Sorgner, S.L. (2020d) Editor's Note. In: *Journal of Posthuman Studies: Philosophy, Media, Technology*, 4(1), 1–4.

Sorgner, S.L. (2020e) Editor's Note. In: *Journal of Posthuman Studies: Philosophy, Media, Technology*, 4(2), 113–118.

Sorgner, S.L. (2020f) Ein europäisches Sozialkreditsystem als pragmatische Notwendigkeit? In: Bauer, M.C./Deinzer, L. (eds) *Bessere Menschen? Technische und ethische Fragen in der transhumanistischen Zukunft*. Springer, Berlin, Heidelberg, 183–200.

Sorgner, S.L. (2020g) *On Transhumanism*. Translated by S. Hawkins, Penn State University Press, University Park, PA.

Sparrow, R. (2007) Procreative Beneficence, Obligation, and Eugenics. In: *Genomics, Society and Policy* 3(3), 43–59.

Sparrow, R. (2011) A Not So New Eugenics. Harris and Savulescu on Human Enhancement. In: *Hastings Center Report* 41(1), 32–42.

Stevens, A. (1994) *Jung*. Oxford University Press, Oxford.

Stiftung Datenschutz (ed) (2017) *Big Data und E-Health*. Erich Schmidt Verlag, Berlin.

Stone, D. (2002) *Breeding Superman, Nietzsche, Race and Eugenics in Edwardian and Interwar Britain*. Liverpool University Press, Liverpool.

Thomsen, M.R./Wamberg, J. (eds) (2020) *The Bloomsbury Handbook of Posthumanism*. Bloomsbury, London.

Thomson, J.J. (1976) Killing, Letting Die, and the Trolley Problem. In: *The Monist* 59: 204–217.

Tönnies, F. (1979) *Gemeinschaft und Gesellschaft*. Grundbegriffe der reinen Soziologie. Wissenschaftliche Buchgesellschaft, Darmstadt.

Tuncel, Y. (ed) (2017) *Nietzsche and Transhumanism. Precursor or Enemy*. Cambridge Scholars Press, Newcastle upon Tyne.

Urpeth, J. (1999) A 'Pessimism of Strength': Nietzsche and the Tragic Sublime. In: Lippitt, J. (ed) *Nietzsche's Futures*. Palgrave Macmillan, London, 129–148.

Vattimo, G. (1997) *Glauben – Philosophieren*. Stuttgart, Stuttgart 1997.

Vita-More, N. (2013) Aesthetics: Bringing the Arts and Design into the Discussion of Transhumanism. In: More, M./Vita-More, N. (eds) *The Transhumanist Reader. Classical and Contemporary Essays on the Science, Technology, and Philosophy of the Human Future*. Wiley-Blackwell, Chichester, 18–27.

Welti, F. (2005) *Behinderung und Rehabilitation im sozialen Rechtsstaat.* Mohr, Tübingen.

Willemsen, R. (2016) *Who We Were. Speech on the Future.* Fischer, Frankfurt a.M.

Woodward, A. (2011) *Understanding Nietzscheanism.* Acumen, Durham, NC.

Young, J. (2006) *Nietzsche's Philosophy of Religion.* Cambridge University Press, Cambridge.

Youngner, S. (ed) (2002) *The Definition of Death: Contemporary Controversies.* Johns Hopkins University Press, Baltimore/London.

Yovel, Y. (1986) Nietzsche and Spinoza: amor fati and amor dei. In: Yovel, Y. (ed) *Nietzsche as Affirmative Thinker.* Martinus Nijhoff Philosophy Library, vol 13. Springer, Dordrecht, 183–203.

Index

References to endnotes show both the page number and the note number (190n34).

23andme 35, 39, 61

A

abortion 158–159
abstraction, cognitive skill of 81
afterlife 8, 74, 116–117, 171, 173, 177, 180, 182
Agar, N. 87, 91
ageing 29, 32, 33–36, 170
alethic nihilism 11, 17–19, 20
algorithms 42, 45, 46, 51–52, 54, 55
'alive,' definition of 26–27, 64, 65
  see also personhood
all-purpose goods 86, 94, 136, 170
Alzheimer's disease 34, 36
Amazon Go 45
angels on a pin argument 10, 22, 176, 186
animals 35, 47, 64, 141–144, 147–149, 151–153, 167
anthropocentricity 142, 145, 151, 153, 186
anti-communitarianism 112–113
anti-utopian transhumanism 162–163, 165–171
apes 34, 142, 148, 149, 151
Apple 46
applied ethics 5
Argentina 150
aristocratic virtues 111–112, 118
Aristotle 28, 86, 89, 129
artificial intelligence (AI) 8, 22, 53, 148, 150, 153, 154–155
artificial reality 24
arts 10–11
asceticism 12
as-good-as-it-gets ethics 15, 38, 46–49, 156, 186
atheism 11, 12, 143, 151
Aueweele, D.V. 11
Augustine, St 123–124
Austria 158
authenticity 136, 137

authoritarianism 19, 190n38
autonomous cars 30, 31
autonomous enhancement 85
autonomy
  carbon-based transhumanism 62–63, 68
  and education 91
  and genetic modification 86, 90–92
  personhood 147, 155, 156
  radical pluralism 136
  use of data 47–49
axolotl genome 35

B

Bacon, F. 37, 39–40
basic needs, meeting 12
basic stable attitudes 50, 89, 120, 128
Belyaev, A.R. 3
Bentham, J. 39
best life 104
  see also good/good life concepts
Beyond Humanism Conference Series 4
big data
  and concepts of harm 160–161
  data protection 30, 37, 42, 43, 53, 57
  expropriation of data 43, 44, 54, 55
  gene technologies 35, 61
  glocalization 57
  health data 32–33, 52, 53
  politics of data collection 31, 36–38
  use of data 9, 33, 37–38, 49, 52, 53
big gene data 30, 99, 170–171
biobags 61
bioengineering 98, 119
bioliberalism 4
biometric data 39, 52
bioprinting 36
bioprivacy 99, 171
Blackford, R. 164
blockchain 30
bodily integrity 43
body modification, right to 5

body monitoring technology 9, 32–33, 36, 160
Bostrom, N. 4, 7, 10, 22, 23, 59, 63, 66, 130–132, 138, 162
brain death 174
brain–computer interfaces 30, 31–32
Buddensiek, F. 89
Buddhism 12, 20, 50, 182

C
Cabat-Zinn, J. 50
Camus, A. 180
cancer 34, 36
carbon dioxide emissions 14–15
carbon-based computers 26
Cartesian duality 64
categorical exceptionalism 25
Chan, S. 72
chimeras 34–35
China 30, 31, 37, 43, 48, 53, 57, 58, 161
chips 9, 23, 31, 32, 42, 160
Christianity 63, 80, 143, 158, 162, 177, 180
citalopram 71, 72, 75
civil war 38
classical ideal see Renaissance ideal
climate change 14–15, 20, 186
cognitive development 78–82, 148
cognitive enhancement 7, 70–71, 82–83
cognitive pain 154, 155, 156
collectivism 58
Colombia 41, 69
communitarianism 109, 111–119, 123, 124, 128
community versus society 111, 113, 114, 124
computer games 10, 23, 146
computer simulation 10, 22, 23–34, 25–27, 29, 182
computer viruses 10, 26, 27, 29, 150, 175
consciousness 10, 20, 27, 64, 152, 154, 156
contact lenses 31
contingent nodal points 18, 19, 21
continual becoming 12, 17–18, 19–20, 47–48, 177, 181
cosmism 3
COVID-19 53–6, 158, 160–161
CRISPR-Cas9 8, 30, 35, 61, 62, 100, 150
critical posthumanism 4, 14, 47, 110, 151–152, 163, 180, 185
Crockett, M. 71, 72
cryonics 6, 170, 172, 173–175
cryptocurrency 30
cultural movement, transhumanism as 2, 3–4
cyborg technologies 9, 23, 32, 176
Cytoplasmic Transfer 69

D
Darwin, C. 2, 65, 66, 91, 112, 126, 137, 142, 147, 149

data 30–60
see also big data
data capital 46
data protection 30, 37, 42, 43, 53, 57
De Chardin, T. 2
De Forest, L. 52
De Grey, A. 34, 129, 170
deafness 87
death 162, 173–174
deceleration 168
definitions of 'transhumanism' 1–2
del Val, J. 138
della Mirandola, P. 117
democracy 17, 22, 30–60, 97, 143
see also liberalism; social-democratic politics
diet 9–10
digital glassess 31
digital immortality 27
digital privacy 22
digitalization 30–60
digitization of personality 10, 25–26
dignity, human 64, 68, 141–147, 149, 151, 153
disability 105–106, 133–134, 135, 136, 147
discrimination 45
'disease' as a concept 6, 85, 94–95
divine spark 149, 150, 151, 186
DNA testing 35, 61
Douglas, T. 71
drives 14, 28, 48, 49, 50, 67, 116, 166
drugs 7–8, 50–51, 71, 72, 75, 77
dualism 63–64, 86, 98, 110, 116, 120, 143, 149, 163
dukkha 12

E
Eastern religious thinking 64
see also Buddhism
economic flourishing 37–38, 58
education
analogy with genetic enhancement 83–99
and autonomy 48
definition of 84, 86
and democracy 17
and the development of talents 120
and equality 93
and parenting 120, 122
reversibility of 89–90
and truthfulness 50
as upgrade 13–14
educational freedom 19, 62, 156
e-governance 55
Eissa, T.L. 79
elections 49
embodied theory of mind 25, 99
embryos 97, 151, 152, 154, 167
enhancement, definition of 83–85

Enlightenment 31, 43, 48, 59, 80, 97, 130, 139, 185
epigenetics 95–96, 106
equality 77, 79, 81–82, 86, 93, 116, 166
Esfandiary, F.M. 2–3
essentialism 20, 64, 151
Estonia 35, 61
eternal recurrence of everything 177–179, 182, 183–184
ethical nihilism 11, 17–19, 20–21, 47, 67
Ettinger, R.C.W. 3, 164
eugenics 15, 84–85, 105, 112, 114–115, 185
evolution
　average life expectancy 34
　carbon-based transhumanism 61, 64–65, 66
　continual becoming 181
　and the development of transhumanism 149
　embodied theory of mind 25
　and epigenetics 96
　naturalism 13, 126
　Nietzschean thinking 112, 113–114, 177
　and the overhuman 113–114
　partner selection 106
　and personhood 148
evolutionary humanism 2, 149
expropriation of data 43, 44, 54, 55
extremism 38

F
Facebook 32, 37, 49
fascism 63, 105, 115
'fat man' trolley problem 72
Faust (Goethe) 179
Ficino, M. 117
Fire, A. 88
firewalls 58, 59, 60
Floridi, L. 40
Flynn, J.R. 81
FM-2030 2–3
Foucault, M. 39
France 158
freedom, norms of 41–42, 79, 139–140, 166, 167
Freud, S. 89, 91
Fuchs, S. 161
Fukuyama, F. 1, 63–64, 149
functional theory of mind 25
Future of Humanity Institute, Oxford 63
Fyodorov, N.F. 3, 164

G
Gattaca 91, 112
gene analysis 9
gene modification 83–99, 106, 170, 176
gene selection 99–108, 119–123
gene technologies
　carbon-based transhumanism 61–108
　and digital technology 30

expanding human boundaries 8–9, 23, 32
　and the health span 170
　moral bioenhancement 77
　prolongation of lifespan 35
　silencing genes 88
　as a vice 110–129
　virtue ethics 119–123
gene therapy 87–89, 94–95
genetic data 35, 39
　see also big gene data
genetic engineering 149, 150
genetic enhancement 119–123
genetically altered living tattoos 26, 36
genome editing 8, 32, 61, 62, 88
Germany
　abortion 158
　biotechnologies 140
　digital data 46, 60
　eugenics 85, 105
　homosexual marriage 41
　human dignity 141–142
　incest 54, 68
　organ donation 56
　personhood 142, 143, 149, 151, 167
　philosophical tradition 62–63, 79, 111, 115, 140, 141–142, 143, 163
　religious freedom 157–158, 159
gesture control systems 32
giftedness 123
Giordano Bruno Foundation 2, 149
globalization 37, 56, 59
glocalization 56–60
Goethe 179
good/good life concepts
　all-purpose goods 86, 94, 136, 170
　anti-utopian transhumanism 166
　common-sense account 132–135
　concepts of human perfection 129–141
　conflicts between differing 137, 142–162
　Enlightenment 80
　ethical nihilism 19
　and freedom 41
　and gene selection 102–103
　and genetic modification 94
　and immortality 165
　increased health span 6, 20–21, 135–138, 139
　liberal ethics 105
　meaning of life 180
　moral bioenhancement 71, 74, 75–76
　Nietzschean thinking 63–70, 114, 118–119, 180
　PB (procreative beneficence) 135
　and prolongation of the health span 6–7, 135–138, 139
　radical pluralism 43, 62, 109, 135–138, 140, 166–167
　rankings of 104, 132

Renaissance ideal 66, 130–132
  and utopias 163
  work-life balance 168
Google 37, 46, 50
Google Glass 31
government harm 158
GPS systems 39
Grof, S. 50
Güth, W. 72

H

Habermas, J. 62, 63, 64, 83, 84, 86, 89, 90,
  91–92, 93, 94, 95, 115, 135–136, 163, 170
hacking 42
Harari, Y.N. 53
harm 18–19, 43, 47, 54, 68, 72–75, 141–162
Harris, J. 72, 78
Hauskeller, M. 7, 162
Hawking, S. 27, 175
Hayles, N.K. 32, 148, 154
health, and human perfection 135–138
health data 32–33, 52, 53
health insurance 43–44, 54, 55, 159
health span
  and ageing 36
  good/good life concepts 6, 20–21,
    135–138, 139
  human-machine interfaces 160
  immortality 169, 183
  and the internet panopticon 42
  prolongation of 6–7, 23, 28, 29, 169–170,
    172, 183
  quality of life 20–21, 28, 29, 43, 59, 159,
    160, 172, 178
health-care 43–44, 56
Heidegger, M. 136, 166
Helbig, S. 163
Heraclitus 12, 138, 179, 183
HIV 16–17, 34, 160–161
Hobbes, T. 40
Hoerster, N. 92
Huawei 58
Hughes, J. 4, 5–6, 12, 19, 50, 64, 130
human dignity 64, 68, 141–147, 149, 151,
  153
human flourishing
  costs of 14
  following psychophysiological
    demands 19, 67
  health span 28
  plurality 18, 41–42, 139
  technology 30, 33, 36
  and total surveillance 38
  and totalitarianism 48
  and utopias 163
human perfection 129–141
human rights 19, 52
human zoo 115

human-animal hybrids 34–35, 61, 68
human-computer interfaces 31–32
humanism 1–2, 4, 147, 151
humanities 83, 98
Humanity Plus 4
human-machine interfaces 30, 31–32, 160
Humby, C. 190n34
humiliation 153–154
Huxley, A. 2, 140
Huxley, A.F. 2
Huxley, J. 1–2, 22–23, 149, 181
Huxley, T.H. 2
hybridization 68, 159–160, 161
  see also human-animal hybrids
hyperautonomy 155–156, 167
hyperhumanism 130
hyperparenting 122, 125

I

ID chips 9
immaterial soul 64, 136, 143
immortality 6–7, 27–29, 164–165, 169, 170,
  171–184
impulse control 49–51, 128
incest 41, 54, 55, 68, 147
individual preferences 137
Inferno (Brown) 10, 63
Institute for Ethics and Emerging
  Technologies (IEET) 4
instrumentalization 92–93
intellectual property 37, 40, 43, 54, 58
intelligence 70, 81, 83, 86, 94, 102, 133,
  135
intelligence explosion 176
Internet of Bodily Things 33, 160
Internet of Things 23, 30, 31, 33, 39, 160
internet panopticon 30, 36, 37–38,
  39–42, 51–52
Invisibile Gorilla test 154
Istvan, Z. 5
IVF
  gene selection 4, 62, 68, 69–70, 77–78,
    100, 102, 121, 123, 127, 129, 170
  gene technologies 8, 9
  moral status of a fertilized egg 69–70
  reproductive freedom 19

J

Japan 35, 61, 68
Japanese tea ceremonies 50
Japlonka, E. 96
jellyfish *Turritopsis dohrnii* 35
Johnson, R. 88
Journal of Evolution and Technologies 4
Journal of Posthuman Studies 4
Journal of Transhumanism 4
judgement 18
Jung, C.C. 171, 184

K

Kahane, G. 99, 100–101, 103, 104, 105, 132, 133, 134, 135
Kant, I. 98, 141, 142–143, 144, 149
Knoepffler, N. 61
Koechy, K. 94
Kurzweil, R. 10, 25, 27, 150, 170, 173
Kuwait 35, 61

L

Lamarckism 96
Lamb, M.J. 96
language 9, 13, 17, 32, 161
Levenson, J.M. 96
liberal eugenics 62, 84–85, 115, 170
liberalism
   autonomy 147
   community versus society 113, 114
   COVID-19 55
   digital data 38, 43, 44, 47
   educational freedom 97
   PA (procreative autonomy) 107
   and paternalism 159
   PB (procreative beneficence) 105
   plurality 128
   Renaissance ideal 130–131
   transhumanists' 5, 19
libertarianism 5, 19, 58, 104
life expectancies 16, 34, 35, 102, 169
life span
   and genetic modification 94
   versus health span 6–7
   increases in 15–16, 26, 28, 34, 173–176
   utopianism 169–170
   see also health span; immortality
literacy 17
living computers 36
living entities, definition of 26–27
living tattoos 26, 36
love, virtue of 123–124
   see also transforming love;
      unconditional love

M

male circumcision 157
Margalit, A. 153–154
Massachusetts Institute of Technology (MIT) 26, 36
master morality 67
masterly virtues 115–118
master-slave types 111, 116
materialism 20
McCaffrey, A.P. 88
meaning of life 110, 163, 171–184
Mello, C. 88
memory, good 133, 135
metahumanism 110, 136, 138, 139

metahumanities 83, 98
microchipping humans see chips
microdosing of drugs 7–8
middle classes 38
Millon, T. 125, 126
Milo, R. 161
mind uploading
   meaning of life 175–176, 181
   personhood 150, 153
   silicon-based transhumanism 22–23, 29
   and the simulation argument 10, 23, 25–27
   utopianism 163, 170, 172
mindfulness 6, 49–51, 128
mirror test 147, 148, 152, 153, 157
mitochondrial disease 69, 84
Moore's law 176
moral bioenhancement 8, 62, 70–83
moral development 78–82, 83
morality 79, 105, 109, 116–117, 141–147, 150
More, M. 3–4, 5, 19, 25, 63, 130, 163
Mormons 163, 180
morphological freedom 5, 19, 138, 156
movies 10, 23, 91
Musk, E. 22, 23–24, 25, 29, 30, 53, 160, 173
Myo 31

N

narcissist personality disorder 125–126
naturalism
   autonomy 91
   good/good life concepts 109
   hermeneutic circle
      naturalism-philosophy 180–183
   immortality 28, 164
   Nietzsche, F. 177
   Nietzschean thinking 64, 65, 126, 142
   nihilism 19–20
   soft naturalism 64, 91
   and transhumanism 11, 12, 13, 17, 149
   and utopias 163
negative freedom
   anti-utopian transhumanism 166
   carbon-based transhumanism 143
   and COVID-19 53
   and gene technologies 62
   and health insurance 159
   philosophical positivity 15
   silicon-based transhumanism 22, 41, 44, 46, 47, 51–52
Neuralink 30, 160
nicotine 50
Niemeyer, C. 65
Nietzsche, F. 3, 6, 12, 20, 62, 63–67, 91, 95–96, 111–119, 123–128, 131, 142, 143, 149, 172, 177, 178, 180, 183

nihilism 11, 17–19, 20, 67–68, 114
non-conscious cognition 154
non-dualism 10, 47, 64, 99, 109–110, 142,
    182, 183–184, 187
non-human animals 35, 47, 64, 141–144,
    147–149, 151–153, 167
non-human cells 161
non-immanence 163
non-violence 168–169

O

Oblomov (Goncharov) 125, 126
OECD (Organisation for Economic
    Co- operation and Development) 58
Olson, K. 52
ontological stability 17–18
open data 58
opt-in systems 55–56
organ donation 56
organic computers 26
overhumans 65, 66, 91, 111–119, 138
oxytocin 76–77

P

PA (procreative autonomy) 100, 107–108,
    156
pain, and personhood 154–155, 156
    see also suffering
panopticon 39
    see also internet panopticon
parahumans 68
    see also human-animal hybrids
parenting 120
    see also education; gene
        selection; hyperparenting
Parkes, G. 64
Parkinson's disease 34, 36
partner selection 106–107, 123
patents 44
paternalism
    and autonomy 47
    community versus society 128
    Enlightenment 48
    good/good life concepts 67, 136, 139
    PB (procreative beneficence) 107
    personhood 143, 151
    relational ethics 185
    religion 158–159
    and utopias 168
    violence of 136, 139
PB (procreative beneficence) 99, 100–108,
    132–135
Pearce, D. 4, 164
personal data, collection of 32–33, 37–38
personhood 47, 110, 140, 141,
    142–156, 167
    see also 'alive,' definition of
perspectivism 18, 113, 173, 181–182

Persson, I. 62, 71, 76, 77, 78, 79, 82, 83
pessimism, philosophical 11–19
PGD *see* IVF
pharmaceutical development 44
pharmacological enhancement 7–8, 50, 170
philosophy of mind 5
Pinker, S. 80, 81, 169, 183
Plato 64, 104, 148, 149, 181
Platonic forms 12, 18
plurality
    building a culture of 167–168
    nihilism 18, 67–68
    philosophical positivity 15
    and power 127
    radical pluralism 68, 109, 135–138,
        139, 166–167
    and reduction of violence 156
    silicon-based transhumanism 38, 42, 43,
        46, 55, 66
    versus totalitarianism 140
politics, transhumanist 5, 19
    see also liberalism; social-democratic politics
population growth 68
positivism 11
positivity, philosophical 13–17
Posthuman Studies 5
posthumanism
    critical posthumanism 4, 14, 47, 110,
        151–152, 163, 180, 185
    and epigenetics 95–96
    gene technologies 29, 91, 150
    history of concept 2–3, 4
    metahumanities 98
    Nietzschean thinking 63, 65
    personhood 142, 156
    philosophical positivity 14
    radical pluralism 137
    and suffering 151–152
    virtues 50
    weak posthumanism 138–139
postmodernism 19–20, 113
post-personhood 154, 155, 167
poverty 11, 16, 20, 45
power
    and data collection 37, 42
    and freedom 45
    glocalization 57
    and the internet panopticon 39–40
    and love 124, 129
    and privacy 40, 51
    will to power 65, 112, 115–118, 126,
        127, 131
predictive maintenance 9–10, 33, 160
primates 34, 142, 148, 149, 151
privacy 22, 40–46, 53, 54, 55
privilege 15
property theory 40, 54
pro-social behaviour promotion 76–77

psychedelic drugs 7–8
psychiatry 89–90
psychophysiological demands 48–49, 67,
 136, 138, 166
psychotherapy 51
psychotropic drugs 50
punishment *see* sanctions

Q
quality of life
 and gene selection 100, 102
 health span 20–21, 28, 29, 43, 59, 159,
  160, 172, 178
 and the internet 40
 and life span 178
 meaning of life 110, 163, 171–184
quantum physics 26, 175, 177

R
racism 45
radical mindfulness 50
radical pluralism 68, 109, 135–138,
 139, 166–167
Rae, M. 34
rape 137, 140, 146
reason 13–14, 149, 155
relational ethics 47–48, 136, 185
relationality 167
religion
 afterlife 8, 74, 116–117, 171, 173, 177,
  180, 182
 Christianity 63, 80, 143, 158, 162,
  177, 180
 customs 157–158
 dualism 63–64
 good/good life concepts 139, 143
 meaning of life 171, 172, 177
 morality 74, 80
 and naturalism 182
 and paternalism 158–159
 personhood 47, 147
 relational ethics 48, 58
 religious transhumanists 182
 and transhumanism 2, 10, 11, 12
 transhumanism seen as
  quasi-religion 163–164
 and utopias 162
Renaissance ideal 66, 67, 117, 130–132,
 134–135
reproductive freedom 19, 54, 63, 70,
 91
RFID chips 9, 23, 33, 45, 160
RNAi mechanism 88
robots 146, 153
Rosa, H. 168
Roser, M. 192n11
Rothblatt, M. 7, 27, 182
Russia 3, 58

S
Salk Institute 35
sanctions 40–41, 43, 44, 46, 54, 55, 74, 157
Sandberg, A. 5
Sandel, M. 109, 112, 119–128
Savulescu, J. 4, 62, 71, 76, 77, 78, 79, 82, 83,
 99, 100–101, 102–108, 129, 131,
 132–135, 136
Schipanski, D. 61
Schopenhauer, A. 11, 12, 20, 180
Schufa 46
Schultz, T. 32
Schuringa, J. 72, 73
science fiction 3, 10, 68
secularism 11
selective attention test 154
self-consciousness 10, 27, 147–148, 149,
 152, 153, 154, 156, 167
self-ownership of body 5
Sender, R. 161
sentience 148, 152, 153, 155, 167
serotonin 72
sex robots 146
Shin, S. 4
shopping 45
silencing genes 88
silicon- versus carbon-based lives 10
simulation argument 10, 22, 23–24, 25–27,
 29, 182
Singer, P. 47, 140, 147–148, 151, 152,
 153, 154–155
siRNA therapy 88
slave moralities 116–117
slavish virtues 117–118, 126–167
Sloterdijk, P. 62, 115
smart cities 23, 30, 36, 161
smartphones 31, 161
social credit 30, 37, 42–46, 48, 52
social media 49
social-democratic politics 5, 19, 43, 44, 58,
 130, 143, 145, 159
Socrates 71, 79, 80, 112
soft naturalism 64, 91
Song, E. 88
Sorgner, S.L. 3, 7, 8, 9, 11, 12, 13, 15, 23,
 30, 32, 36, 41, 47, 57, 61, 64, 65, 67, 70,
 71, 74, 79, 80, 83, 84, 85, 87, 91, 106,
 112, 113, 115, 116, 118, 119, 122, 123,
 124, 126, 128, 138, 139, 141, 143, 149,
 151, 161, 166, 167, 168, 170, 172, 176,
 177, 181
soul 25, 64, 117, 136, 143, 148, 182
space travel 68
Spain 41, 55, 68
speciesism 47, 151
Spinoza, B. 20, 179
Star Trek 154, 156, 167
steered organisms 9, 13, 32, 161

Stiftung Datenschutz 30
Stoics 179
Strugatsky, A.N. 3
Strugatsky, B.N. 3
suffering 12, 13, 14, 20, 147, 148,
    151, 154–156
    see also health span
suicide 28
superman 3, 6, 109, 138
surgery, as enhancement technique 7
surveillance 22, 30, 38, 39–42, 55
survival of the fittest 66, 137, 150
sustainability 15
Sweatt, J.D. 96
Sweden 42

T
telos 120
'The Land of Cockaygne' 162
Thiel, P. 5
thought-based computer interfaces 32
three biological parents 61, 68, 69, 84
three person marriage 69
time 148–149, 180
Tolcapone 51
Tönnies, F. 111
total surveillance 22, 32, 38, 41, 42, 44, 45,
    51–52, 55, 57
totalitarianism 40, 47, 48, 67, 76, 123, 140,
    157, 168
Transcendence (Pfister) 10, 23, 63, 153
transforming love 121, 124–125
Transhumanist Party (US) 5
TransVision conference 4
Trolley Problem 72
Trump, D. 58
truth 181–182
truthfulness 49–51, 110, 112, 116, 127, 129

U
Ultimatum Game 72
unconditional love 109, 112, 121, 123, 124,
    125–127, 129
Underwear Bomber 72, 73
US
    data collection by private companies 42,
        43, 57, 61

digital data 58
three biological parents 69
utilitarianism 5–6, 39, 62, 72, 101, 104, 117,
    131, 147
utopianism 7, 15, 62, 110, 140, 157,
    162–171

V
vacation time 168
vaccinations 9, 43, 44, 72–73
Vattimo, G. 18, 21
violence
    as-good-as-it-gets solutions 46–49
    and concepts of the good 67
    and harm 156–157
    and legal use of force 73
    moral development 80–81
    nihilism 18, 21
    of paternalism 136, 139
    PB (procreative beneficence) 104–105
    social credit 45
virtual reality 23–27
virtue ethics 5, 111–129
virtue of justice 77–78
virtues 49–51
viruses 27
Vita-More, N. 3
voice interfaces 32
VPNs (virtual private networks) 59

W
Wagner, R. 124, 129
wearable computers 31
will to power 65, 112, 115–118, 126,
    127, 131
work-life balance 168
World Transhumanist Association (WTA) 4
Wright, W. 52

Y
Young, J. 111

Z
Zamyatin, Y.I. 3
Zarathustra 65, 66
zebra fishes 68

Printed and bound by CPI Group (UK) Ltd, Croydon, CR0 4YY

27/10/2024

14580559-0001